CW00433565

Social Economy of the Metropolis

"The world [is] a progressively realized community of interpretation"

Josiah Royce (1913: 394)

Social Economy of the Metropolis

Cognitive-Cultural Capitalism and the Global Resurgence of Cities

Allen J. Scott

OXFORD
UNIVERSITY PRESS

OXFORD

UNIVERSITY PRESS

Great Clarendon Street, Oxford OX2 6DP

Oxford University Press is a department of the University of Oxford.
It furthers the University's objective of excellence in research, scholarship,
and education by publishing worldwide in

Oxford New York

Auckland Cape Town Dar es Salaam Hong Kong Karachi
Kuala Lumpur Madrid Melbourne Mexico City Nairobi
New Delhi Shanghai Taipei Toronto

With offices in

Argentina Austria Brazil Chile Czech Republic France Greece
Guatemala Hungary Italy Japan Poland Portugal Singapore
South Korea Switzerland Thailand Turkey Ukraine Vietnam

Oxford is a registered trade mark of Oxford University Press
in the UK and in certain other countries

Published in the United States
by Oxford University Press Inc., New York

© Allen J. Scott, 2008

The moral rights of the author have been asserted
Database right Oxford University Press (maker)

First published 2008

All rights reserved. No part of this publication may be reproduced,
stored in a retrieval system, or transmitted, in any form or by any means,
without the prior permission in writing of Oxford University Press,
or as expressly permitted by law, or under terms agreed with the appropriate
reprographics rights organization. Enquiries concerning reproduction
outside the scope of the above should be sent to the Rights Department,
Oxford University Press, at the address above

You must not circulate this book in any other binding or cover
and you must impose the same condition on any acquirer

British Library Cataloguing in Publication Data

Data available

Library of Congress Cataloging in Publication Data

Scott, Allen John.
  Social economy of the metropolis: cognitive-cultural capitalism and the
global resurgence of cities / Allen J. Scott.
      p. cm.
   ISBN 978-0-19-954930-6
1. Sociology, Urban.   2. Cities and towns—Growth.   3. Capitalism.   I. Title.
   HT153.S59 2008
   307.76—dc22   2008016910

Typeset by SPI Publisher Services, India
Printed on acid-free paper

ISBN 978-0-19-954930-6

Preface

Many if not most social scientists today would probably be ready to acknowledge the broad claim that capitalism somehow or other paves the way for large-scale urban growth and development. Beyond this vague putative consensus, however, urban studies at the present time is a cacophony of divergence and disagreement, understandably so, perhaps, in view of the fact that cities are implicated in virtually every dimension of modern life. My purpose in this book is not to range across this entire domain of academic discussion, much less to provide a comprehensive review of the current state of the field. I propose, rather, to argue in favor of a few basic principles that I believe provide some critical foundations for refocusing urban theory on the essential nature of urbanization in capitalism and then to deploy these principles in an investigation of urban growth and development patterns in the current conjuncture.

My point of departure in the latter task resides in the general claim alluded to above, and specifically in the rather more controversial idea that modern cities are *in the first instance* dependent expressions of the logic and dynamics of the wider economic environment. This point of departure must immediately be qualified by reference to a second basic premise about the nature of urbanization, for cities are not just simple, passive excrescences of the capitalist economy. To the contrary, they also play an intrinsically active role in the unfolding of the economic order, both directly and indirectly (via diverse extra-economic conditions of existence). In the absence of urbanization, capitalism as we know it would be a very much more low-key affair, if it could exist at all. In other words, yes, modern cities make their historical appearance and develop as a consequence of capitalist economic dynamics, but they are also critical moments in the continuation of capitalism as a going concern. This dual role of cities is, of course, an expression of the wider patterns of circular and cumulative causation that underlie the formation of virtually the entire space-economy of the modern world.

The intimate recursive connection between urbanization and capitalist society at large means that every historical version of capitalism is associated with distinctive types of cities, and vice versa. The first machine age in the nineteenth century was founded on a peculiar kind of urbanization composed of tightly knit agglomerations of factories and workshops intertwined with dense tracts of cheap housing where the largely impoverished proletariat eked out a living. The second machine age, with its core technological-*cum*-organizational bases residing in mass production and large-scale growth poles, was typified by the formation of extended metropolitan areas and by a dominant division of urban social space into white-collar and blue-collar neighborhoods, itself a reflection of the basic division of labor in fordist society. The third machine age, which began to emerge some time toward the late 1970s and early 1980s, is now bringing forth a number of startlingly novel forms of economic and social organization based in significant ways on new technologies of computation and information, and cities are once again responding to this state of affairs in their double role as both outcomes and fountainheads of economic development. The current moment, I shall argue, is one in which a specifically *cognitive-cultural* economy has made its historical appearance, with profound consequences for the configuration of contemporary urbanization and urban social life. Even as the cognitive-cultural economy proliferates across the globe, cities represent the crucibles in which the new economic order is being forged and in which many of its essential bases are assembled into local socioeconomic systems.

The present book is an effort to put some conceptual and descriptive order around these different claims. The argument proceeds on three main registers. First, I examine some of the essentials of urban theory generally, with the broad objective of rearticulating the urban question in a way that is relevant to today's world, and where by the term "urban question" I mean an explicit fusion of a scientific problematic and a political project directed, on the one side, to the comprehension, and on the other side, to the progressive reconstruction of urban life. I argue that while the urban question quite properly changes its shape and form from time to time depending on conjunctural circumstances, it is nevertheless durably rooted in a systematic logic that reflects the tension-filled dynamics of urban space. Second, I offer a variety of theoretical and empirical observations about the functional characteristics of today's cognitive-cultural economy as manifest in sectors like technology-intensive production, financial and business operations, fashion-oriented manufacturing, cultural industries, personal services, and so on. These sectors are growing with great rapidity in the world's largest cities at the present time, and they play an important role in the great urban resurgence that has occurred over the last few decades all across the globe. Third, I explore the detailed spatial ramifications of the cognitive-cultural economy in contemporary cities and the ways in which it intersects with many other kinds of urban processes. In particular, the cognitive-cultural economy appears to be ushering in major shifts in the division of labor and social stratification in capitalist society, as marked by a growing divide between a privileged elite stratum of workers (managers, professionals, technicians, etc.) on the one side, and a kind of new lumpenproletariat on the other, and this state of affairs, of course, is pregnant with consequences for the social life of cities. In all of this, I lay special emphasis on the idea of the *social economy* of the metropolis, which is to say, a view of the urban organism as an intertwined system of social and economic life played out through the arena of geographic space.

I should add that as the investigation proceeds, empirical points of reference are for the most part confined to North American cities, but frequent allusions are also made to cities in other parts of the world, and in the final analysis, my purpose is to provide a fairly comprehensive synthesis of the issues. With this end in mind, the penultimate chapter of the book seeks to throw a sharp light on the ways in which urbanization in general is today caught up in a series of complex interrelations between the cognitive-cultural economy and processes of globalization. One striking expression of this state of affairs can be observed in the emergence of selected city-regions as economic and cultural flagships of the new world order. I am concerned in this book not only to provide a few guidelines as to how we might approach these matters but also to work out a number of suggestions about the political issues and tasks of reform that are becoming increasingly urgent in large cities consequent upon the rise of a globalizing cognitive-cultural economy.

Even so, I am conscious that the basic terrain of analysis under scrutiny here is extraordinarily complex and conceptually challenging and that my attempt to deal with it is provisional at best. I recognize, too, that my stress on the economic dimensions of urbanization—even if the economic is constantly qualified by reference to the social—will not be to the taste of those numerous scholars today who tend to emphasize more culturally inflected forms of urban analysis. My objective is not to deny the significance of cultural variables in the shaping of modern cities, much less to treat them as residual consequences of economic life. I am more than ready to acknowledge that culture as construed by latter-day culture theorists has many far-reaching effects on the economy in general and has notably powerful impacts on how cities function and look. Indeed, the quotation from Royce that stands at the head of this book points already to this admission, for one of my concluding propositions is that economy and culture appear to be converging together into new and peculiar structures of meaning whose focal points are the great city-regions of the global era. All that being said, I am nevertheless intent in what follows on an attempt to assert the essential genesis of contemporary forms of urbanization in the capitalist economic order, an objective that I seek further to justify and to realize in Chapters 1 and 2. It is my hope that the modest contribution to urban analysis put forth here will help to spark off further extended research and debate, for the issues to which I allude are of some conceptual significance, and, in more concrete terms, they refer to a number of major practical challenges for cities and society at large in the looming era of cognitive-cultural capitalism.

Part of the material of the book has been previously published in the following journals: *Annals of Regional Science, European Planning Studies, Internal Journal of Urban and Regional Research, Journal of Urban Affairs, Social Forces, Urban Affairs Quarterly,* and *Urban Studies.* I am grateful to the publishers of these journals for permission to reproduce extended passages here. However, all of the material reproduced in this book has been significantly edited, recombined, and rewritten for the purposes of the present volume, and a considerable amount of new content has been added. The argument as a whole, therefore, is very much more complex and goes significantly beyond the already published work in which it has its roots. I also wish to express my gratitude to the National Science Foundation (specifically for grant number BCS-0749038), the Richard S. Ziman Center for Real Estate at UCLA, and the Committee on Research of the Academic Senate also at UCLA for research funding that has enabled me to carry out detailed investigations of a number of the basic themes that are explored below. Finally, I thank my friends and colleagues David Rigby and Michael Storper for their helpful comments on early drafts of sections of this book.

AJS

Contents

List of Illustrations

List of Tables

1

The Resurgent City

A salient feature of societies in which capitalistic rules of order prevail is that they are invariably marked by high and usually increasing rates of urbanization. This relationship between capitalism and urbanization has been manifest since the first stirrings of modern industrialization in eighteenth-century Britain, and it has continued down to the present day as country after country has been caught up in the expansionary thrust of capitalism across the globe. The relationship is neither a mere contingency nor an instance of a simple progression from cause (the economy) to effect (urbanization). The city is not only a response to the pressures of capitalism (via the formation of distinctive clusters of capital and labor on the landscape) but also a basic condition of the continued social reproduction of the capitalist economic system as a whole (Scott 1988, 1993, 1998*b*).

Equally, the city is something very considerably more than just an economic phenomenon, in the sense of a workaday world of production and exchange activities, for it is also a congeries of many other kinds of social relationships (including a very definite political component). Capitalism is from top to bottom caught up in an extraordinarily complex network of socioeconomic interdependencies. In the sphere of the urban, moreover, these interdependencies take on a peculiar significance and interest, for the city is one of those sites where the social and the economic are most visibly connected together, above all as they are projected through the dimension of urban space. That said, I want to propose right at the outset that there is a sense in which economic production and exchange activities play a particularly privileged—though by no means exclusive—role in the actual emergence of modern cities. In the absence of this dimension and above all in the absence of its powers of bringing on locational agglomeration, cities would probably be little more than minor aggregations of communal life. Certainly, the absence of this dimension would tend drastically to undercut any possibility that sustained and localized increasing-returns effects might make their appearance, and hence by the same token to dampen the spirals of circular and cumulative causation that we commonly associate with large-scale urban growth today. The urban behemoths of the contemporary world—places like New York, Los Angeles, London, Paris, Tokyo, São Paulo, Mexico City, Bombay, Hong Kong, Singapore, and so on—could surely never have arisen had there not existed a powerful capitalistic dynamics pushing them ever forward as foci of national and, nowadays indeed of, global economic life.

Urbanization in modern society, however, is never a smoothly operating process, partly because of the wayward course of wider national and international affairs, partly because cities are always susceptible to the buildup of internal disruptions and social collisions. Over the period of the long post-War boom in the United States and Western Europe, large industrial cities flourished on the basis of a dynamic fordist mass-production system with its voracious demands for direct and indirect material inputs and its dependence on enormous local reservoirs of labor. But just as cities thrive when their economies are expanding, so also do they enter into crisis in the converse case. By the mid-1970s, many of the cities that had benefited most from the fordist system were brought to the verge of bankruptcy, as mass production entered a protracted period of adversity exacerbated by foreign competition, labor-management discord, and stagflation, and as rapidly shifting production technologies steadily eroded the economic sustainability of the old order. As a consequence, the 1970s and 1980s represented a period in which many analysts published strongly pessimistic accounts of the future of cities and regions, and in which notions of long-run secular decline were very much in the air. Yet this was also a period in which the seeds of an unprecedented urban renaissance were being planted, as expressed in both accelerating shifts toward a more knowledge-based economy in primate cities like New York, London, and Paris and in the emergence of new industrial spaces and communities in many formerly peripheral areas throughout the developed and the less-developed world.

Even as this renaissance was occurring, pessimism about the future of cities remained at a high pitch in many quarters, though the diagnosis was now taking a very different turn from that of the economic declinism of the 1970s and 1980s. Over the 1990s, as consciousness about the potentialities of new communications technologies gathered momentum, it was proclaimed in several quarters that distance was effectively dead and that a new era of globally deconcentrated interaction was about to be ushered in (Cairncross 1997; O'Brien 1992); it was accordingly thought by many that cities would henceforth steadily lose much of their reason for being and that a trend toward massive population dispersal was about to begin. It goes without saying that new communication technologies have vastly extended our powers of interaction across geographic space, and have accordingly brought distant communities closer together, but the predicted process of urban decline has thus far failed signally to show up in statistical data (see Table 1.1). To the contrary, mounting empirical evidence and theoretical argument point to the conclusion that globalization and its expression in a virtual space of flows is actually intensifying the growth and spread of cities throughout the world (cf. Hall 2001; Taylor 2005). This trend is all the more emphatic because of the major sea changes that have been proceeding of late, not only in basic technologies but also in the organizational and human-capital foundations of contemporary capitalism. These changes mobilize, to a hitherto unprecedented extent, the knowledge, cultural assets, and human sensibilities of the labor force in the production of ever increasing quantities of "intellectual property" and other forms of congealed intelligence, information, and affect.

I propose, in what follows, to deal with these and related issues by reference to the idea of a new cognitive-cultural economy. The locational foundations of this new economy reside preeminently in large metropolitan areas. Concomitantly, and as Cheshire (2006) has suggested, cities in the early twenty-first century exhibit strong symptoms of resurgence, especially in comparison with the dark days of the dying fordist regime. Exploration of this complex terrain of relationships involves three principal lines of argument. One of these comprises an effort *passim* to rethink the scope and substance of urban theory generally. Another attempts to provide a detailed empirical and analytical account of the current conjuncture—marked as it is by both radical shifts in the nature of capitalist enterprise and an intensifying trend to globalization. Yet another seeks to work out some of the more important implications of these shifts for an

Table 1.1. The world's 30 largest metropolitan areas ranked by population (in millions) in the year 2005

Metropolitan area	Country	1975	2005	2015 (predicted)
Tokyo	Japan	26.6	35.2	35.5
Mexico City	Mexico	10.7	19.4	21.6
New York	USA	15.9	18.7	19.9
São Paulo	Brazil	9.6	18.3	20.5
Mumbai	India	7.1	18.2	21.9
Delhi	India	4.4	15.0	18.6
Shanghai	China	7.3	14.5	17.2
Calcutta	India	7.9	14.3	17.0
Jakarta	Indonesia	4.8	13.2	16.8
Buenos Aires	Argentina	8.7	12.6	13.4
Dhaka	Bangladesh	2.2	12.4	16.8
Los Angeles	USA	8.9	12.3	13.1
Karachi	Pakistan	4.0	11.6	15.2
Rio de Janeiro	Brazil	7.6	11.5	12.8
Osaka-Kobe	Japan	9.8	11.3	11.3
Cairo	Egypt	6.4	11.1	13.1
Lagos	Nigeria	1.9	10.9	16.1
Beijing	China	6.0	10.7	12.9
Manila	Philippines	5.0	10.7	12.9
Moscow	Russian Federation	7.6	10.7	11.0
Paris	France	8.6	9.8	9.9
Istanbul	Turkey	3.6	9.7	11.2
Seoul	Republic of Korea	6.8	9.6	9.5
Chicago	USA	7.2	8.8	9.5
London	United Kingdom	7.5	8.5	8.6
Guangzhou	China	2.7	8.4	10.4
Bogotá	Colombia	3.1	7.7	8.9
Tehran	Iran	4.3	7.3	8.4
Shenzhen	China	0.3	7.2	9.0
Lima	Peru	3.7	7.2	8.0

Source: United Nations (2006).

understanding of the changing form and functions of cities, including their internal spatial arrangements and their collective political order. The rest of this chapter is devoted to a preview of these arguments.

The Socio-Geographic Constitution of the City

Presumably, few urban analysts would disagree with the notion that the city is a distinctive spatial phenomenon embedded in society, and therefore expressing in its internal organization something of the wider social and property relations that characterize the whole. Any attempt to define the city in more concrete terms, however, is almost certainly liable to generate considerable controversy. A cursory examination of the literature on urbanization reveals a cacophony of interests, perspectives, and points of empirical emphasis that are all said to be urban in one way or another. Empirical phenomena are regularly qualified as being urban for no more obvious reason than that their spatial limits coincide more or less with the limits of the city. Much of the time, as well, cities are simply equated with "modern society" as a whole. However, there can be no simple equation to the effect that if, say, 80 percent of the population of the United States lives in metropolitan areas, then cities must represent 80 percent of everything that constitutes American society. Education, for example, or for that matter, ethnicity, fashion trends, or crime are not *intrinsically* urban issues, even though there might be senses—even important senses—in the second instance, in which we can say that they intersect with an urban process. If we are to make sense of this confusion (and in order to understand just exactly what it is that is resurgent, and why), we need some sort of problematic, that is, a circle of concepts by which we might pinpoint a social logic and dynamics that clearly demarcate the urban within the wider context of social life at large.

A basic point of departure here is the observation that one of the things all modern cities share in common is their status as dense polarized or multipolarized systems of interrelated locations and land uses. No matter what other social or economic peculiarities may be found in any given instance, cities are always sites or places where many different activities and events exist in close relational and geographic proximity to one another. I reaffirm this truism at the outset because I want to argue that this is the point of departure for any theory that seeks to capture the intrinsic, as opposed to the contingent, features of urbanization. Proximity and its reflection in accessibility is an essential condition for effective unfolding of the detailed forms of interdependence that constitute the lifeblood of the city, and that are all the more insistent in the world of modern capitalism with its finely grained divisions of labor. In turn, the competitive quest for proximity on the part of diverse economic and social agents brings into being an intra-urban land market that results in powerful processes of locational sorting so that different parts of the city come

5

to be marked by different specialized types of land use. The same processes induce the piling up of diverse activities at selected points of high gravitational intensity, with the greatest density invariably occurring at and around the very center of the city. The complex, evolving whole constitutes what I have referred to in earlier work as the *urban land nexus* (Scott 1980); though as it stands here, the concept remains something of a formal skeleton devoid of social content. Accordingly, we now proceed to probe the meaning of this notion further in the context of three main questions. First, what is it in general that drives the search for proximity? Second, and as a corollary, what is it specifically that constitutes the central function or functions of the city as such? Third, what administrative and political tasks are intrinsically conjured up as the logic of intra-urban space unfolds? The answers to these questions provide important clues about the mainsprings of the resurgent city.

We can think of many reasons why large numbers of people would want to participate in spatially agglomerated activity systems. One widely cited factor is the search for some kind of human and cultural community; another is the efficiencies that can be obtained by building many different kinds of social and physical infrastructures in compact local settings (Glaeser and Gottlieb 2006). Factors like these no doubt make some contribution to the overall process of urbanization, but their powers of centripetal attraction must surely become exhausted long before we arrive at the kind of large metropolitan areas that are found in capitalist society. As an initial *argumentum ad hominem*, it seems hard to imagine that the massive urban growth that has occurred in the more economically advanced societies over the last couple of centuries might be ascribed simply to some sort of communal impulse or to the indivisibilities of infrastructural artifacts. In any case, there can be no sustained process of urban development in the absence of employment opportunities for the mass of the citizenry. The same opportunities, moreover, are socially constructed within the synergistic networks of productive capital (industrial, service, retail, etc.) that express the many different lines of mutual interdependence and interaction knitting individual units of economic activity in the city into a functioning system. These networks function not only as inert sources of jobs for the populace but also as dynamic social mechanisms, much given to expansionary thrusts, and, on occasions, to overall contraction (with inevitably negative consequences for the rest of the urban complex). Thus, large cities can always be represented as huge axes of production and work that function primarily on the basis of their interrelated firms and their dense local labor markets. As we shall see, the inherent economic dynamism of these systems is underpinned by the propensities for learning, creativity, and innovation that frequently characterize thick grids of human interaction. These complex phenomena constitute the fundamental engine of urban growth and development, providing, of course, that market outlets for final products can be found. The workings of this engine generate powerful agglomeration economies that set up a strong gravitational field, and therefore, as the engine is mobilized, the city expands by continually drawing in new additions to its stock of capital and labor. Cities are also increasingly enmeshed in processes of globalization, but this does not mean, as Amin and Thrift (2002) suggest, that they therefore cease to function as sites of local interdependency and economic power. On the contrary, the more the urban economy is able to reach out to distant markets, the more it is able to grow and differentiate internally, leading in turn to reinforcement of its agglomerative magnetism. To be sure, countervailing trends to decentralization are also always at work, but processes of urban expansion have thus far—with only occasional and temporary interruptions—tended to outrun any long-term tendency to decline.

Precisely because the city is not just an inert aggregate of economic activities, but is also a field of emergent effects, it is by the same token a collectivity in the sense that the whole is very much greater than the sum of the parts or, more to the point, its destiny is in important respects shaped by the joint outcomes that are one of the essential features of urbanization as such. These effects are evident in the guise of negative and positive externalities, agglomeration economies, localized competitive advantages, and so on. They constitute a sort of commons that is owned by none but whose benefits and costs are differentially absorbed by sundry private parties, oftentimes unconsciously so. In the absence of clearly defined property rights, the commons is resistant to market order, and without the intervention of some rationalizing agency of collective decision-making and control is liable to serious problems in regard to the ways in which its benefits and costs are produced and spread out over urban space. This means in turn that there is an intrinsically positive social role for agencies of policy implementation and planning in the city with a mandate to seek out solutions to the problems posed by the commons in all its complexity. These agencies sometimes exist at extra-urban levels of institutional organization, though the principle of subsidiarity suggests that they will usually be constituted as integral elements of urban society as such. Their role consists, both theoretically and practically, in many-sided efforts in the production space and the social space of the city to provide beneficial public goods, to enhance the supply of positive externalities, to bring negative externalities and other urban breakdowns under control, and to ensure that rewarding opportunities which would otherwise fail to materialize are pursued as far as feasible. The city is also a place where latent political contestation and collisions about the use and allocation of urban space are always present, and from time to time these tensions break out in open conflict. These tensions are a further constituent of the urban land nexus and contribute significantly to the overall managerial problems that it poses. Indeed, they are often sparked off by collective interventions that seek to impose remedial order in the urban system but then generate the need for further rounds of intervention in order to deal with the reactions of the citizenry to the initial effort of remediation.

It is in this broad context that we need to situate any claims about the resurgence of the city. The remarks outlined above suggest that it will be fruitful to approach this issue with a focus on questions of production and social reproduction combined with a clear sense of the imbrication of these phenomena in the geographic logic of the city and an insistence on the intrinsically collective nature of the dynamics of intra-urban space. In order to set the scene further and to fix ideas, we take up the story with a brief rehearsal of urban problems and predicaments during the fordist episode of capitalist development. This may at first appear as something of a diversion, but its relevance will become more sharply apparent as we see how it throws light not only on the widespread urban crisis that preceded the current period of urban resurgence but also on the general problematic of urbanization sketched out above.

From Growth to Crisis in Fordist Mass-Production Society

Over much of the twentieth century, the dominant (though by no means exclusive) model of economic growth and development in North America and Western Europe revolved around the mechanisms of large-scale mass production (Coriat 1979; Piore and Sabel 1984). This system of economic activity was based on capital-intensive lead plants linked to lower tiers of direct and indirect input suppliers, thus forming dynamic growth poles in industries like cars, machinery, domestic appliances, electrical equipment, and so on (Perroux 1961). Many individual producers caught up in these growth poles had a strong proclivity to gather together in geographic space, and the resulting industrial clusters constituted the backbones of large and flourishing metropolitan areas. The production system itself was distinguished by a fundamental twofold division of labor comprising blue-collar workers on the one side and white-collar workers on the other. This division of labor was then cast out, as it were, into urban space where it became reexpressed, imperfectly but unmistakably, as a division of residential neighborhoods, upon which was superimposed a further pattern of social segmentation based on differences of race and national origins. In the United States, the Manufacturing Belt was the main locus of this peculiar form of industrial-urban development. The equivalent area in Western Europe consisted of a swath of territory stretching from the British Midlands through northeastern France, Belgium, southern Holland, and the Ruhr, with outliers in northwest Italy and southern Sweden.

The core regions of the mass-production economy expanded rapidly over the middle decades of the twentieth century, and they developed apace as new investments were ploughed into productive use and as streams of migrants converged upon the main metropolitan areas. Notwithstanding persistent decentralization of routinized branch plants to low-wage locations in the periphery, the core regions continued to function as the main foci of national economic growth, for as the arguments of Myrdal (1959) and Hirschman (1958) make clear, the synergies or increasing-returns effects generated within the major cities of the mass-production system kept them consistently in positions of economic leadership relative to the rest of the national territory. Moreover, from the New Deal of the 1930s onwards, mass-production society was subject to ever more elaborate policy measures designed to maintain prosperity and social well-being. After World War II, these measures evolved into the full-blown keynesian welfare-statist policy system designed to curb the cyclical excesses of the mass-production economy and to establish a safety net that would help to maintain the physical and social capacities of the labor force, especially in periods of prolonged unemployment. The scene was now set for the long post-War boom over the 1950s and 1960s, and for the climactic period of growth of the large cities that functioned as the hubs of the mass-production economy. This policy system was supervised and controlled by central governments, but, as Brenner (2004) has argued, it was in many important ways put into effect through municipal agencies, and it had major transformative impacts on urban space. Throughout the 1950s and 1960s, urban renewal, housing programs, intra-urban expressway construction, suburban expansion, and diverse welfare schemes performed the interrelated functions of maintaining economic growth and keeping the urban foci of the boom operating in a reasonably efficient and socially manageable way.

By the early 1970s, the classical mass-production system in North America and Western Europe was beginning to shows signs of stress, and as the decade wore on it entered into a long-run period of exhaustion and restructuring. The reasons underlying this development involve many different factors including changes in technologies, the rise of superior versions of classical mass production in Japan and elsewhere, and big shifts in national and international market structures. The details of these changes need not detain us here, except to note that the endemic pattern of decentralization of production units away from core areas was by the early 1970s turning into a rout, and the formerly thriving industrial cities of the system were now faced with massive job loss, unemployment, and decay. Deindustrialization of the old manufacturing regions advanced at a swift pace over the 1970s, and with the deepening of the crisis the US Manufacturing Belt itself came more commonly to be known as the Rustbelt. In the metropolitan areas that had formerly functioned as the quintessential centers of the long post-War boom, the watchwords now became stagnation and decline, most especially in inner city areas where residual working-class neighborhoods were marked by a pervasive syndrome of unemployment, poverty, and dereliction. In the United States, even those metropolitan regions that had weathered the economic crisis relatively well were left with deeply scarred central cities as a result of industrial decentralization and restructuring. By this time, too, much of the job flight that was occurring was no longer simply directed to national peripheries but was increasingly aimed at low-wage locations in the wider global periphery.

Analysts such as Blackaby (1978), Bluestone and Harrison (1982), Carney et al. (1980), and Massey (1984) now began to write notably gloomy accounts about the outlook for the cities and regions that had most benefited from economic growth over the period of the long post-War boom. For many of these analysts, any prospect of a vigorous urban recovery seemed to be extremely dim indeed. The neoliberal political agenda initiated by Reagan in the United States and Thatcher in the United Kingdom confirmed this pessimism in many quarters, especially as much of the scaffolding of the keynesian welfare-statist system was now being steadily dismantled, and as more and more stable high-paying, blue-collar jobs continued to disappear permanently from the urban scene.

Into the Twenty-First Century (1): Cities and the New Economy

Some time in the late 1970s, at the very moment when this gloom seemed to be reaching its peak, intimations of an alternative model of economic organization and development started to appear in various places. Several attempts to conceptualize this model have been offered under the rubric of "sunrise industries," or "flexible specialization," or "post-fordism," or the "knowledge economy," or "cognitive capitalism," or simply the "new economy" (cf. Bagnasco 1977; Esser et al. 1996; Markusen, Hall, and Glasmeier 1986; Piore and Sabel 1984; Rullani 2000). Right from the start of this effort of conceptualization, many analysts noted that a spurt of agglomeration and urbanization seemed to be following on the heels of the new model, especially in regions that had been bypassed by the main waves of industrialization in the immediate post-War decades, such as the US Sunbelt or the Third Italy, as well as in selected densely urbanized areas like the New York metropolitan area or the London region that had in any case always had a relatively diverse economic base.

There was, and is, much debate about the character and meaning of the new economy that began to emerge some two or three decades ago (see, e.g., Gertler 1988; Hyman 1991; Pollert 1991; Sayer 1989; Schoenberger 1989). Whatever specific controversies may be at stake in this regard, there does not seem to be much disagreement about the fact that a rather distinctive group of sectors much typified by deroutinized production processes and relatively open-ended working practices began to move steadily to the fore of economic

11

development after the early 1980s (though strictly speaking, the roots of the new economy can be traced back to the 1960s and even to the 1950s if we consider such early precursors as Hollywood after the 1948 Paramount Decree or Silicon Valley after the mid-1950s). The core sectors of the new economy include technology-intensive manufacturing, services of all kinds (business, financial, and personal), cultural-products industries (such as media, film, music, and tourism), and neo-artisanal design- and fashion-oriented forms of production such as clothing, furniture, or jewelry. These and allied industries have now supplanted much of the mass-production apparatus as the main foci of growth and innovation in the leading centers of world capitalism where they constitute the main sectoral foundations of what I referred to above as a new cognitive-cultural economy. On the occupational side, this phenomenon has been accompanied by the formation of a thick stratum of high-wage professional and quasi-professional workers concerned with tasks that can be seized in generic terms as scientific and technological research, administration and deal-making, representation and transacting, project management and guidance, conception and design, image creation and entertainment, and so on. These elite occupational activities are at the same time complemented by and organically interrelated with a second stratum composed of poorly-paid and generally subordinate workers engaged in either manual labor (as e.g. in apparel manufacturing or in the assembly of high-technology components) or low-grade service functions (such as office maintenance, the hospitality industry, childcare, janitorial work, and so on). While the tasks faced by workers in the lower tier are often quite routine and monotonous, there is even here a tendency—especially in large US cities—for many of them to require a substantial degree of performative flexibility and judgment and/or cultural sensitivity on the part of employees (McDowell, Batnitzky, and Dyer 2007).

The cognitive-cultural economy, then, is marked by the increasingly flexible and malleable systems of production (with their ever-varying palette of goods and services) that are now so strongly present at the leading edges of the contemporary economy. As it happens, the cognitive-cultural economy is also highly concentrated in urban areas, and many of its most dynamic segments have a particular affinity for major global city-regions like New York, Los Angeles, London, Paris, Tokyo, and so on (Daniels 1995; Krätke and Taylor 2004; Pratt 1997; Sassen 1994; Taylor 2005). The reasons for the attraction of cognitive-cultural industries to locations in the city reside primarily in the organizational logic of the new economy generally, in combination with the ways in which the uncertainties that loom over these industries are moderated by the size and density of the urban milieu. The cognitive-cultural economy is focused on small production runs—even in large firms—and niche-marketing and its core sectors tend to be radically deroutinized and destandardized. Individual producers are almost always caught up in detailed transactions-intensive networks of exchange and interdependence with many other producers, often in situations where considerable interpersonal contact is necessary for successful mediation of their common affairs. These networks, in addition, are susceptible to much instability as firms adjust their process and product configurations and hence swing continually from one set of input specifications to another. Equally, local labor markets are subject to a great deal of unpredictability as a consequence of the volatility of production activities and the growth of temporary, part-time, and freelance forms of employment, even among well-paid and highly skilled workers (Angel 1991; Blair, Grey, and Randle 2001). These features of the cognitive-cultural economy alone are calculated to encourage a significant degree of locational convergence of individual producers and workers in selected urban areas, not only as a way of reducing the spatial costs of their mutual interactions but also as an instrument allowing them to exploit the increasing-returns effects that flow from the risk-reducing character of large aggregations of latent opportunities. However, there is a further factor that contributes greatly to this process of convergence. As interacting firms and workers gather together in one place, and as auxiliary processes of urban development are set in motion, a distinctive field of creative and innovative energies is brought into being in the sense that the links and nodes of the entire organism begin to function as a complex ever-shifting communications system characterized by massive interpersonal contacts and exchanges of information (Scott 2006b; Storper and Venables 2004). Much of the information that circulates in this manner is no doubt little more than random noise. Some of it, however, is occasionally of direct use to the receiver, and, perhaps more importantly, individual bits of it—both tacit and explicit—combine together in ways that sometimes stimulate the formation of new insights and sensibilities about production processes, product design, markets, and so on. In this manner, strong creative-field effects may be mobilized across sections of intra-urban space, leading to many individually small scale but cumulatively significant processes of learning and innovation within any given locality.

Into the Twenty-First Century (2): *Urbs et Orbis*

The contemporary resurgence of cities is inscribed in and greatly magnified by a deepening trend to globalization. This trend is expressed in the vast geographic extension of the range of markets that any given city can reach and in the ever-deepening international streams of labor (both skilled and unskilled) that pour into the world's most dynamic metropolitan regions.

As these trends unfold, the geographic pattern and logic of globalization itself is shaped and reshaped in various ways. In the old core–periphery model of world development, the advanced capitalist countries, and especially their major metropolitan areas, were often seen as being essentially parasitic on the cheap labor of the periphery by reason of unequal development and exchange (Amin 1973; Emmanuel 1969). A major attempt to update the model was made by Fröbel et al. (1980) in their theory of the new international division of labor, where they claimed that the core tends to develop as a specialized center of white-collar work (command, control, R&D, etc.), while the periphery evolves as a vast repository of standardized blue-collar work. None of these different claims stands up very well in confrontation with the specifics of urbanization and globalization over the last couple of decades, not so much because they tended to overestimate the nature of exploitation in capitalism, but rather because they failed radically (understandably enough in view of their vintage) to assess the subtleties of geographical eventuation in a world of intensifying international interaction. In particular, the rise of the new economy with its associated underbelly of sweatshops and low-grade service activities employing huge numbers of unskilled immigrant workers has meant that major cities of the core are now directly interpenetrated by growing Third World enclaves, while many parts of the erstwhile periphery have become leading foci of high-skill technology-intensive production, business and financial services, and creative industries. To be sure, we can still detect important elements of the core-periphery model in the great expansion of international labor outsourcing from high-wage to low-wage countries that has been occurring over the last couple of decades (Gereffi 1995; Schmitz 2007). In spite of these continuing echoes, much of the old core-periphery pattern of international economic development seems to be subject to gradual supercession by an alternative geographic structure comprising a global mosaic of resurgent cities that function increasingly as economic motors and political actors on the world stage. Not all of these cities participate equally in the cognitive-cultural economy, though all are tied together in world-encircling relations of competition and collaboration; and those that have emerged or are emerging as leaders in the cognitive-cultural economy function to ever greater and greater degree as the cynosures *par excellence* of the contemporary global system.

In the context of these developments, the resurgent cities and city-regions of today's world are evidently beginning to acquire a degree of economic and political autonomy that would have been for the most part unimaginable in the earlier fordist era when the national economy and the nation state represented the twin facets of a dominant sovereign framework of social order and political authority. In line with the general spatial rescaling of economy and society that has been occurring as globalization runs its course, something like a new regionalism is also becoming increasingly discernible. Thus, just as individual identities, social being, and institutional structures are increasingly subject to reconstitution at diverse scales of spatial resolution, cities and city-regions are now starting to play a role as important economic and political components of the world system. In view of this remark, the early speculations of Jacobs (1969) about cities (in contrast to states) being organically fitted to serve as units of functional economic organization and social life must be seen as having been remarkably prescient. If anything, the waning of keynesian welfare-statism and the turn to devolution in the context of an insistent focus on markets and competitiveness has helped to bring the substance of these speculations closer to concrete reality. Major city-regions are everywhere struggling with a multitude of social experiments as they attempt to consolidate their competitive advantages in the face of the deepening predicaments posed by globalization, and as they search out local institutional arrangements capable of responding effectively to idiosyncratic local economic needs and purposes. In an era of intensifying neoliberalism and globalization, when national governments are less and less able or willing to cater to every regional or sectional interest within their

15

jurisdictions, cities must now either take the initiative in building the bases of their own competitiveness and social stability or face the negative consequences of inaction. One noteworthy expression of this trend—especially in large global city-regions—is the growing realization that some sort of administrative and institutional coordination across the urban land nexus as whole is a necessary condition for achieving overall efficiency, workability, and local competitive advantage. The force of this realization is such as to have encouraged diverse experiments in the consolidation of local institutional arrangements in many different places, including the creation, or proposals for creating, cross-border metropolitan governance structures, as in the Øresund region in Scandinavia, the Pearl River Delta in southeast China, or, more fancifully, perhaps, Cascadia in the western US–Canadian borderlands.

Life and Politics in the Resurgent City

It is clear that the resurgent city of the contemporary era presents several radical points of contrast with the fordist industrial metropolis of the mid-twentieth century. These contrasts are manifest in both the economic bases of these two categories of city and in their general social structure. Moreover, while each type of city exhibits significant racial and ethnic diversity, today's resurgent city, certainly in North America, is probably marked by more cultural variety than at any time in the past, and, more crucially, is increasingly a magnet for immigrants from both developed and less developed countries all over the world. Even the upper tier of the workforce in resurgent cities contains significant and increasingly higher proportions of immigrants from other countries. The net result is a new sort of cosmopolitanism in the populations of these cities (Binnie et al. 2006), not so much the rarified cosmopolitanism of an earlier era whose defining feature was its implicit allusion to a free-floating group of individuals of dubious origins but elite pretensions, but an everyday cosmopolitanism that freely accepts an eclectic mix of urban identities and cultures as a perfectly normal aspect of modern life.

Just over a century ago, Simmel (1903/1959) characterized the denizens of the modern city as a mass of mechanistically interconnected but psychologically disconnected individuals. Much of this characterization is no doubt still valid in the context of the resurgent city, with its synchronized in-step rhythms of work and its atomized forms of social life. The possessive individualism of urban society has, if anything, made considerable headway by comparison with the cities of middle and high modernity. There is much evidence to suggest that traditional urban or neighborhood webs of community and solidarity continue to disintegrate while norms of market order and meritocratic criteria of human evaluation penetrate ever more deeply into the fabric of social existence. Even the apparently countervailing expansion of civil society—NGOs, nonprofit organizations, philanthropic foundations, and the like—might be taken as a sign of underlying processes of social fragmentation and the retreat of formal governance mechanisms than it does of political solidarity and mobilization (Mayer 2003). Still, the new kinds of consumerism and hedonistic social rituals of contemporary urban life offer consolations of sorts in the face of what Simmel calls the "unrelenting hardness" of cities, at least for privileged segments of society. Lloyd and Clark (2001) have alluded to something of what I am reaching for here with their description of the modern metropolis as an "entertainment machine," that is, as a place in which selected spaces are given over to ingestion of the urban spectacle, upscale shopping experiences, entertainment and distraction, nighttime scenes, and occasional cultural adventures in museums, art galleries, concert halls, and so on. These spaces dovetail smoothly in both formal and functional terms with the gentrified residential neighborhoods and high-design production spaces that are the privileged preserve of the upper tier of the labor force of the modern cognitive-cultural economy.

At the same time, life and work in the resurgent city are subject to high levels of risk, both for lower-tier and for upper-tier workers. As social welfare provisions are steadily pared away and as traditional union organization declines in contemporary society, lower-tier workers in particular are exposed to the full stresses and strains of this situation, most notably those who make up the large and increasing corpus of marginalized (often undocumented) immigrant workers. This is a world, however, in which the possibilities of large-scale political mobilization seem more and more remote, and in which collective action on the part of municipal authorities seems increasingly to assume the mantle of professionalized, technocratic agency lying outside the sphere of agonistic political encounters. By the same token, much of the intra-urban conflict over the welfare and distributional impacts of planning action that was so characteristic in the past (and that reverberated especially throughout the working-class neighborhoods of the fordist city) has now more or less subsided into the background. In some respects, the only resonances that remain of the disappearing atmosphere of open political contestation in the large metropolis emanate from the identity-based claims and conflicts that seem now largely to have displaced popular agitation in regard to economic justice. Even in its currently depoliticized form, however, collective action in the resurgent city is far from being a merely neutral or disinterested force. Municipal authorities today are acutely focused on the concerns of property owners and business, and virtually everywhere are engaged in schemes directed to the shoring up of local competitive advantages and the attraction of inward investors. Large-scale redevelopment projects, expenditures on urban amenities, city-marketing, the promotion of local festivals, and so on figure prominently among these kinds of schemes. Of course, the endemic tensions of urban life still have a proclivity to spring forth into spontaneous open conflict. The point can be dramatically exemplified by reference to the Los Angeles riots of 1992, as well as to the disturbances that broke out in the immigrant quarters of the Paris suburbs in the latter part of 2005 and that then spilled over into other parts of the metropolitan area. The paradox of the resurgent city is the escalating contrast between its surface glitter and its underlying squalor.

2

Inside the City

Urbanization and the Urban Question

In the previous chapter I sketched out a broad overview of some of the major socioeconomic forces that mold the spatial form and evolutionary trajectory of urban areas, and I alluded to a basic conception of the city as a locus of densely polarized and interdependent locational activities. In this context, I made special reference to the deep urban crisis of the 1970s and the subsequent resurgence of cities after the early 1980s. We now need to look with considerably more care and detail at the internal constitution of cities, and in this manner to build a foundation for the subsequent investigations of urban fortunes in the twenty-first century. This exercise involves an attempt to rearticulate the urban question in contemporary capitalism at large and in cognitive-cultural capitalism more particularly. My motivation for this line of attack comes in part from what I take to be a growing loss of focus in much that currently passes for urban analysis, and from a dissatisfaction with the increasingly frequent conflation of social issues in general with urban issues in particular. A pertinent point in this regard, as Cochrane (2007) has suggested, is the apparently endemic confusion about just how the domain of urban policymaking is constituted and how it might be distinguished—if at all—from nonurban policymaking. Some attempt to clarify this confusion is important not only in its own right but also as a guide to strategic mobilization in the interests of urban reform.

I proceed at the outset by restressing the ontological status of the city in capitalist society as an agglomerated system of multifarious phenomena (transport facilities, factories, offices, shops, houses, workers, families, ethnic groups, and so on) integrated into a functional whole by a dominant process of production and accumulation.

This system is energized by myriad individual decisions and actions coordinated via market mechanisms; but it is also—and of necessity—a major site of collective coordination and policy intervention. What imbues this many-sided grid of activities with a distinctively *urban* character cannot be discovered by focusing attention on the aliquot empirical relata that make it up, but only by investigating their peculiar form of social and spatial integration, that is, the variable geometry of their expression as interrelated socio-geographic outcomes (land uses, locational patterns, spatial structures, and so on) jointly organized around a common center and associated subcenters of gravity. I shall argue, as well, that while it may be possible to identify a minimal urban problematic that is more or less applicable across the history and geography of capitalism, cities are nonetheless subject to marked conjunctural peculiarities reflecting changes in the wider social and economic context. The investigation of correspondingly localized urban questions in time and space therefore also constitutes a crucial research moment.

Over the twentieth century, many different views of the urban question have been on offer, and each of these has typically been deeply colored by the peculiar circumstances of history and place in which it was formulated. In the 1920s, the Chicago School of urban sociology put forward what subsequently came to be a leading concept of the city based to a significant degree on the status of Chicago as a center of large-scale immigration from many different European countries and as a social cauldron that Upton Sinclair had characterized as a "jungle." Not surprisingly, the Chicago School approach combined a powerful sense of the massive growth of the large industrial metropolis with a Darwinist conception of the struggle between different social groups for living space (Park, Burgess, and McKenzie 1925). In the late 1960s and early 1970s this hegemonic account was challenged both implicitly and explicitly by seminal urban analysts like Castells (1972), Harvey (1973), and Lefebvre (1970) who were then engaged in the codification of what was rapidly to become a widely acclaimed notion of the city within the broader theory of political economy, one that focused specifically on urban outcomes within a broad web of fordist production relations and keynesian welfare-statist policy arrangements. At the core of these new descriptions of the urban, despite their individual differences, lay a concern with the disparities and injustices of "urban society" and with the unequal socio-spatial allocation of the collective consumption goods (capital-intensive infrastructures, public housing, educational facilities, etc.) that compose much of the physical groundwork of modern cities. Above all, the city was seen as a site of basic distributional struggles played out through public investment and planning activities in the built environment, and an arena in which issues of social justice and the democratic right to urban space were continually at stake. The so-called Los Angeles School of urban analysis (see, e.g., Scott and Soja 1996) was in many ways both a prolongation of this earlier mode of investigation and—in part at least—an anticipation of a further set of conceptual developments concerned with issues of everyday life, social identity, and urban culture (see, e.g., Amin and Thrift 2002; McDowell 1999; Soja 2000; Watson and Gibson 1995). Dear's account of the "postmodern urban condition" captures much of the spirit of the latter trend (Dear 2000).

All of these approaches to urban theory and the urban question offer diverse insights into how cities work in the advanced capitalist world, though none, I believe, provides a sufficient framework for an understanding of the essential mainsprings of the urban process at the beginning of the twenty-first century. While the achievement of any such understanding obviously requires the work of many hands, the present exercise is a modest attempt to push the discussion forward by means of an inquiry into the general structure and dynamics of intra-urban space and its concrete forms of expression at the present time. In this endeavor, I hope to capture something of a latent synthesis based on a reformulated political–economic approach to urbanization together with an explicit concern for the sociocultural dynamics at work in the unfolding of life inside the city. This synthesis is colored by three major overarching developments in the contemporary world. First, of course, a new cognitive-cultural economy has come steadily to the fore and is now giving rise to major rounds of growth and internal social differentiation in the world's large metropolitan areas. Second, an overriding turn to neoliberalism in governmental policy stances in many of the more advanced capitalist countries has ushered in a climate of increasing fiscal austerity, and is associated, among other things, with massive public withdrawal from all forms of redistributive policy, both national and local. Third, globalization is advancing apace, bringing cities all over the world into new configurations of competition and collaboration with one another, and at the same time stimulating many different experiments with new forms of institutional response at the local level. These three points are essential to any reconsideration of urbanization and the urban question in the current conjunture, both because of their implications for the character of urbanization in the sense given above and because they betoken a number of profound shifts in the geography and balance of political power in contemporary society generally. My objective, then, is not only to recover a specifically urban problematic but also to pinpoint something of the nature of the urban question *qua* scientific undertaking and political project relevant to the present moment in history.

Urban Space: Private and Public Dimensions

In the year 2004, 73.6 percent of the 293.6 million residents of the United States lived in metropolitan areas of 250,000 or more. With such a massive absolute and relative concentration of population in the country's largest cities, it is tempting to think that the urban encompasses virtually the totality of social life, and in fact, this slippage occurs repeatedly in both the academic and popular literature on urban affairs. I shall argue, however, that the urban, which is assuredly a social phenomenon, is nonetheless something very different from society as a whole, and that if we are to make sense of its internal logic we must distinguish unambiguously between that which is merely contingently urban and that which is intrinsically so. In short, and to echo a now largely forgotten refrain originally expressed by Castells (1968) the urban, if it has any sense at all, must be carefully distinguished as an object of inquiry from society at large.

The need for this distinction is easy to state in principle, but making it clear is extraordinarily difficult in practice. Raymond Williams (1976) says that "culture" is one of the two or three most difficult words in the English language, but I would add that "urban" must surely also rank close to the top. There is perhaps a natural tendency in any attempt to identify a phenomenon as complex, multifaceted, changeable, and omnipresent as the city, to curtail the search for some sort of overall characterization and to seize on those empirical features that are currently most obviously in view in terms of both their empirical weight and political implications (ethnicity, for example, or gender, or social conflict). Still, some baseline point of departure is eminently desirable as a way of sorting out the essentially urban properties of the endless substantive contents of the city. My own starting position here is to ask: what minimal definition provides us with a useful analytical purchase on the phenomenon of the urban while being able to accommodate its numerous empirical variations in space and time (though my references to space and time will be confined in the present context to the geography and history of capitalism)? With this standard of performance in mind, I suggest that we initiate the argument with a provisional three-tiered concept of the urban as (a) a dense assemblage of social and economic phenomena (of which units of capital and labor are of primary importance) organized around a common spatial center and associated subcenters of gravity, (b) tied together both directly and indirectly in relations of functional interdependence (interfirm input–output relations, the journey to work, interindividual networks of various sorts, and so on), and (c) forming a systematically differentiated arrangement of spaces or land uses.

I shall elaborate upon this rather bare characterization of the urban with very much more conceptual and empirical detail at a later stage in the discussion. What is essential for now is that this basic definition already commits us to an intrinsically spatial concept of the form and function of the city in the concrete context of capitalist economic, social, and property relations. Hence, a given event or process, such as industrial production, technological research, ethnic differentiation, crime, or education, is relevant to urban analysis to the degree that it makes a difference in terms of the kind of spatial structure identified above. Curriculum changes in elementary schools are not much likely to be of direct relevance to an urban problematic in my sense, but the allocation of schools to neighborhoods is unquestionably so. Certainly, other possible perspectives of the urban are conceivable in principle and evident in practice— not least, perhaps, the semiotics/poetics of the city as celebrated by writers like Aragon, Baudelaire, or Benjamin—but the particular formulation offered here is of particular interest and significance because it codifies in distilled form the roots of a unique syndrome of interconnected social outcomes and political dilemmas (cf. Vigar, Graham, and Healey 2005). Note that I refrain in this discussion from any engagement with one of the more common but surely one of the least interesting problems posed in the quest for a definition of the urban, namely, how and where should the boundaries of the city be drawn? In functional terms, the city's gravitational field extends asymptotically outward across the whole of geographic space, which suggests, in fact, that our definition is really a subset of a wider set of issues concerning society and space at large, and therefore must ultimately be generalizable to include *inter*city relations as well. In other words, questions about urban space are in the end subsidiary elements of an overall spatial problematic. In view of this observation, the best course of action to follow when practical delimitation of an urban area is required (e.g. for statistical purposes) is no doubt simply to follow established practice, which is to ignore the pseudo-problem of the "real" boundaries of the city, and to settle for some convenient administrative or governmental unit.

The latter point suggests, indeed, that our initial definition is still not quite as pregnant as it might be, given its silence in regard to any sort of governance, policymaking, or planning activity relevant to the city. The sphere of intra-urban space is constantly subject to direct and indirect policy interventions by many different tiers of government, from the municipal through county and state to the federal level. Sometimes these interventions are addressed to quintessentially urban issues in the meaning already adumbrated, as in the case of urban renewal programs or local economic development initiatives. On other occasions, they may have a hybrid character in that they have both explicitly urban and nonurban components, as illustrated by certain aspects of keynesian welfare-statist policy in the post-War decades (see below). For the rest, much policymaking activity, especially at the federal level, has no directly urban objective in my sense, but has important secondary impacts on the city. In fact, there are few public policies or actions of any kind that do not have some ultimate urban effect. This is especially so given that local governments function not only as arrangements for dealing with purely internal problems in their jurisdictions but also as administrative devices for relaying national and state policy down to the subnational level. In these circumstances, we may ask, what is urban policy *as such*, and does it make sense to attempt to distinguish it from the wider policy environment? Cochrane (2007) tends to the view that no plausible lines of demarcation can be established in this regard. While this view has much to commend it, the problem still remains that we must build into any viable conception of the urban its status not just as a domain of market outcomes based on individual decision-making and action but also as an organic collectivity that poses a variety of administrative and political dilemmas that must be addressed if city life is not to implode in upon itself (Scott 1980).

In one sense we can answer the question posed above by saying that urban public policy is simply policy directed to the urban as defined. This way of handling the issue, however, evades a more acute part of the question, namely, what is it *in the nature of cities* (as distinct from society as a whole) that generates public policy imperatives and that shapes the substantive content of policymaking activities? From the perspective of hard-core neoclassical theory this latter question is for the most part nugatory, for in an ideal market individual decision-making and behavior alone will ensure a Pareto-efficient equilibrium outcome. The urban arena, however, is structurally and chronically resistant to general competitive equilibrium, not only because of disruptions due to market failure in the narrow technical sense but also because the latent synergies, political tensions, and social break-downs that reside in intra-urban space call forcefully for remedial collective decision-making and action. The viability of the city, in terms of efficiency, workability, and livability, depends therefore on the existence of policymaking infrastructures capable of carrying out corrective programs of intervention and regulation. These infrastructures may be constituted by a diversity of governmental and non-governmental institutional forms (including public–private partnerships) though their modes of operation always reflect the structure of underlying urban realities. The logic of urbanization itself generates collective action imperatives and imposes definite constraints on the potential achievements of any such action, but public regulation of the urban sphere is also shaped by political pressures reflecting the interests and political objectives of various social constituencies in the city. These remarks, ultimately, are echoes of the general principle that public policy, like urbanization, is a concrete social phenomenon and is therefore comprehensible only in relation to the pressures and possibilities that characterize the circumstances out of which it springs, including the governance and collective action capacities of society as a whole.

This essentially social-realist view of the policy process, even at this initial stage of discussion, goes against the grain of certain mainstream theoretical advocacies to the effect that policymaking— whether addressed to urban issues or not—can best be understood as a predominantly procedural exercise in pursuit of abstracted norma-tive goals, and whose powers of accomplishment depend primarily on turbocharging the policy apparatus itself, rather like the "eight-step path of policy analysis" proposed by Bardach (1996). There are, of course, procedural and normative elements in all attempts to make and implement policy. That said, and irrespective of the potential organizational or technical failures that Bardach is concerned to correct, the policy process is always organically embedded in a wider social and political milieu that fundamentally shapes its substantive content and meaning, even if we can sometimes only seize the logic of these matters in *a posteriori* terms. Similarly, the specialized aims and goals of urban planning (as distinct from the broader concept of urban policy generally) can best be understood as a set of socially and politically determinate practices directed to the remediation of particular forms of urban dysfunctionality, above all in the domains of land and property development. Equally, conceptions of urban planning that radically abstract the planner from the realities of everyday practice in concrete urban situations are necessarily either radically unfinished or vacuous as descriptions of what urban planners actually do and what they can realistically accomplish in their professional engagements (Roweis 1981). These kinds of conceptions are rife in much of planning theory where they assume such guises as the identification of planning with the search for "rational-comprehensive" solutions to urban problems, or as a means of reaching toward some socially decontextualized idea of the "good city," or as an exercise in hermeneutics or social empowerment, or even as a reflection of the psychic constitution of planners. Hooper (1998), for example, writes of planning in nineteenth century Paris as "a masculinist fantasy of control," a formulation that patently fails to deal with the vastly more central issues revolving around the mounting economic and political problems of central Paris in the mid-nineteenth century, and the social imperative of reordering the internal space of the city in response to the pressures of modernization and economic growth. Of course, none of this discussion is intended to depreciate the role of political mobilization in the shaping of policy decisions, or to suggest that particular planning solutions necessarily match up in a one-to-one way with given urban problems. The point rather is that forms of mobilization and the outcomes that they help to bring about are always grounded in a complex urban and social reality that simultaneously engenders problems calling for solution and offers diverse opportunities for concrete intervention, but also imposes numerous constraints on what can be accomplished. This is precisely why political action must always be accompanied by and shaped by disciplined analysis of the salient concrete situation.

Economy, Society, and Urbanization

Urbanization, as we know it, is a tangible expression of the social and property relations of capitalism, though cities are by no means simply microcosms of society as a whole, and they are marked by many unique and distinctive emergent effects. With this proviso in mind, I now want to develop further the claim that the genesis and evolution of cities can be traced ultimately to the logic of production, exchange, and associated forms of social reproduction in capitalism. This opening salvo rather bluntly and unfashionably puts heavy emphasis on the economic as the motive force underlying contemporary urbanization processes. I propose now to argue more vigorously on behalf of this initial line of emphasis, first of all by providing a highly generalized account of what I shall call *proto-urban forms* and their origins in processes of locational agglomeration, and then second of all by showing how this account takes on historical and geographical specificity in the context of contemporary economic realities.

Certainly, and as the enormous body of published research on urban and regional development testifies, the bare logic of locational agglomeration in capitalism can be very largely accounted for in terms of basic economic dynamics (though even here, as I shall indicate, a number of critical social and political variables intrude on the manner in which these processes operate). The proto-urban outcomes that are the initial spatial expression of these dynamics represent the main drivers of localized economic growth. My argument suggests, by implication, that in the absence of this primary economic dimension, cities would in all probability be little more than, say, rudimentary service hubs, or simple aggregations of like-minded individuals, or specialized centers of administrative activity, but in any case, strictly limited in size and overall complexity by comparison with large metropolitan areas of today. As it is, the dense spatial concentration of human activity that is the essence of the modern city can principally be ascribed to two mutually reinforcing moments of genesis rooted in the economic order. One involves the locational clustering of many different but interrelated units of capital and labor as a strategy for reducing the spatially dependent costs of their joint interactions, both traded and untraded. The other resides in the increasing-returns effects that are set in motion as clustering proceeds and that endow the emerging agglomeration with multiple competitive advantages and social benefits (and, it may be noted for future reference, with a circular and cumulative pattern of growth) (see, among others, Cooke and Morgan 1998; Duranton and Puga 2004; Scott 2006*b*; Storper 1997).

These moments of genesis emerge out of a complex tissue of economic relationships, which, as they work their way through geographic space, lead to the formation of diverse clusters of economic activity on the landscape. Among these relationships, we may identify three at the outset that occur in various combinations and intensities in different intra-urban activity systems, that is

- *The networks of specialized but complementary units of production that typically lie at the functional core of any urban area of significant size.* The traded and untraded interdependencies that run through these networks ensure that selected groups of interlinked producers will tend to seek out locations that lie in close proximity to one another. To be sure, there may well be instances where the individual production units in any given city are entirely unrelated to one another, though such instances would seem to be few and far between. Rather, what we more commonly observe in cities that are more than, say, just cheap labor pools for a few branch plants is at least a core congeries of producers tied together in various kinds of transactional relationships. Often enough, several distinctive but frequently overlapping congeries of this type (including, by extension, retail and personal service activities) can be found in any single city.

- *The multifaceted local labor markets that tie the production space and the social space of the city together into a functioning whole.* These local labor markets are typically composed of interconnected subsystems linking different functional and spatial segments of the urban economy. Useful information about job opportunities and employment conditions tends to circulate with ease through these markets, and this phenomenon constitutes one of the important sources of increasing-returns effects in the urban economy. The intra-urban process of socialization and habituation of the labor force are also of major significance and are often tinged with distinctive place-specific attributes.

- *The learning and innovation effects that almost always emanate from the numerous socioeconomic interactions that occur within the local production system and its associated labor market.* In fact, the city as a whole functions as a sort of creative field—albeit one that is also completely open to the rest of the world—in which multiple bits of information flow with special intensity between the diverse units of economic and social activity contained in urban space. As this occurs, different individuals or teams of individuals may acquire novel insights of various sorts, and although any concomitant innovations may be individually of only minor proportions, in cumulative terms this process can result in important forward advances in the overall competitive prowess of the local economy.

Economic activities with high levels of interdependence along these three axes often have a propensity to cluster together in geographic space, though they only begin to assume the form of something approaching a full-blown city as various noneconomic phenomena coalesce both functionally and spatially around them and restructure them. All the same, locational convergence is strongly marked at this proto-urban stage because of the spatially dependent costs that are incurred as these activities interact with one another and as the spatially dependent benefits that are generated as a function of their mutual proximity start to materialize. The force of this convergence is all the greater where the system is subject to uncertainty and instability, for individual levels of risk can often be greatly reduced where conditions of sociospatial aggregation prevail. For example, if a sufficient number of firms of a particular type cluster together in geographic space, a highly specialized subcontractor offering services that are only occasionally in demand may now find it possible to survive in the same locality. Convergence is yet further compounded by the savings that can be gained by concentrating large-scale infrastructural investments in a relatively limited number of areas, and by the emergence of institutional arrangements designed to regulate aspects of the local economy that are susceptible to market failure and other social irrationalities. The transactional efficiencies and increasing-returns effects (more specifically, agglomeration economies) generated in these ways continually buttress one another and establish the conditions under which processes of cumulative causation come into being so that as firms and workers mass together, yet further massing ensues, and so on, in successive rounds of temporal interdependence. If external markets are expanding (and if alternative sources of production are not deeply threatening) growth of the city will continue indefinitely and the internal complexity and dynamism of the local production system will tend to become more and more robust. The appearance of negative externalities will from time to time put limits on this process of growth, but urban planners and policymakers are also continually at work in attempts to mitigate the worst effects of these impedances. Under conditions of advancing globalization, the productive clusters that come into being in this way function increasingly as nodes within a far-flung network of competitive and collaborative relations, and in which comparative advantage as a reflection of natural endowments becomes progressively overridden by socially and politically constructed competitive advantages rooted in the logic of urbanization itself.

Once all this has been said, the idea of the city in any more complete meaning of the term can only be finally realized after we add to this initial formulation a further series of social and political variables that act back upon and reshape the diverse phenomena of proto-urban space. The constitution of social life is of the first importance in this regard, for individuals play a critical role not only as workers in production space but also as actors within the social space of the city. In particular, workers are caught up in a residential/domestic milieu, and, more broadly, in processes of social reproduction that help among other things to sustain their embodied human capital. This situation is obviously of great complexity, though in at least some degree, it would appear to reflect various needs and preferences that flow from workers' positions in the division of labor. Lefebvre (1974: 41) expresses this idea in the following terms:

[Urban] space contains more or less appropriately located *social relations of reproduction*, that is, bio-physiological relations between the sexes and different age groups in the specific context of the family—and *relations of production*, that is, the division of labor and its organization, and hence hierarchized social functions. These two sets of relationships, production and reproduction, cannot be separated: the division of labor is reflected and sustained in the family; conversely, family organization influences the division of labor.

In the absence of magic carpets, this twofold process of production and social reproduction must be played out within the spatial compass of a feasible daily activity system. This means that the intra-urban production and social spaces of the city are of necessity tightly interwoven with one another. The two spaces are then selectively integrated together by local labor market processes and commuting patterns, which in their turn are sustained by infrastructural networks that potentiate mobility and interconnection (Graham and Marvin 2001). Moreover, just as the production space of the city is susceptible to internal differentiation, so too is social space, which consistently decomposes into distinctive neighborhoods, some of them reflecting divisions of labor in the urban economy (e.g. neighborhoods made up mainly of intellectual workers versus neighborhoods made up mainly of manual workers), some of them rooted in other dimensions of social fragmentation (e.g. race, ethnicity, or religion).

The continuing pervasiveness of socio-spatial segmentation in urban areas is testimony to the essential and ever-increasing human diversity of cities in contemporary society. I say essential here because much of this diversity can be traced back directly to the fundamental developmental trajectory of cities in capitalism and the labor market dynamics that ensue. On the demand side of the process, expanding metropolitan areas are almost always unable to satisfy their labor market needs by means of internal demographic growth, so that deficits can only be made up by inward flows of migrants. On the supply side, both skilled and unskilled workers in large cities increasingly originate in far-flung parts of the world. Low-wage workers from relatively underdeveloped areas are especially attracted to major metropolitan areas in wealthier countries because of the voracious demand for cheap labor by the low-grade workshop, factory, and service activities that are integral to the modern urban economy. Accordingly, wherever we find poverty and its associated misfortunes—which is to say, above all, in the world periphery— there we almost always observe outward streams of migrants directed to large urban areas all over the globe, and prepared to work in them at the most menial tasks available. These migrants typically account for much of the ethnic and cultural diversity that is to be found in urban areas today, and the trend has intensified in recent decades as barriers to international travel have declined. Moreover, as one type of minority group in the metropolis becomes assimilated through upward mobility into mainstream society, so other minorities from other parts of the world move in, leading to continual renewal and intensification of urban social and spatial fragmentation. Rates of assimilation vary greatly from one minority to another, however, depending on both the sociocultural assets specific to the minority itself and the prejudices-*cum*-rigidities of the wider host society. In American cities, these processes of intra-urban social differentiation are further complicated by the presence of significant

African-American minorities who are legitimate citizens but subject to significant negative discrimination (see Chapter 6).

The discussion thus far identifies two of the basic elements that constitute the inside of the city, namely a proto-urban production space, and an associated social space (with subjacent spaces devoted to transport, shopping, leisure, etc.). A third analytical maneuver is now called for in order to bring this material into a reasonably full portrayal of the city as a whole. This involves explicit consideration of the collective order of the city and the formation of pertinent institutions of governance and coordination. As a preliminary to this maneuver we need to reemphasize and reexpress the concept of the city as a specifically *geographical* phenomenon—that is, as a dense spatial fabric of economic and social *relata* tied together and structured by their mutual interdependencies. As we have observed, these interdependencies also involve multiple externalities, increasing-returns effects, agglomeration economies, and other social benefits (and costs) that are produced and consumed by all individual participants in the urban system but that lie outside the framework of individual ownership rights and market exchange. To this degree, the production and allocation of these phenomena is devoid of any self-organizing optimizing rationality, whether social or economic, and can therefore best be rationalized by means of policy intervention. The city is thus liable to persistent inefficiency and failure in the absence of internalization by the collectivity. As a consequence, the urban commons intrinsically emerges as a permanent and powerful attribute of intra-urban space. It is precisely this status of the modern city as *res publica* that now brings us back to the question of public policy and planning as necessary constituents of the urban process *in the strict sense* in contemporary society.

Collective Order and Policy Imperatives in the City

As the central function of accumulation and its associated processes of social reproduction are projected through the medium of urban space they assume peculiar tangible forms of expression, and evoke equally peculiar forms of policy attention. A more conventional way of making essentially the same point is to say that cities are arenas within which multiple opportunities are always and inevitably available for public effort to shore up the efficiency and workability of urban society as a whole. My focus here is on public policy specifically targeted to the management and reordering of urban space as identified above. It goes without saying that any attempt to draw a strict line between urban and nonurban policies still remains a rather thankless task. However, in view of all that has gone before, we would doubtless have little hesitation in consigning, say, federal deficit-reduction measures to the domain of the nonurban (notwithstanding the asseverations of President Clinton's National Urban Policy Report (HUD 1995)). Deficit reduction reflects the play of practical circumstances and political debates that lie for the greater part outside the realm of the urban as identified, even if it has many secondary and tertiary urban impacts. By contrast, we would surely acknowledge that, say, legislation regarding community investment banks, suburban sprawl, or the construction of rapid-transit systems is immediately and intrinsically an element of the urban process in capitalism, for these cases are bound up directly with definite articulations of urban space. Notwithstanding this (approximate) distinction between urban and nonurban public policy, we do need to keep in mind two important provisos as expressed earlier. The first is that there is no reason in principle why urban policy, as such, cannot flow from institutions of governance at many different levels of scale, and not just the local (Uitermark 2005). The second is that public policy can be very much a hybrid affair that operates in both urban and nonurban dimensions simultaneously. We shall encounter a dramatic case of this kind of hybridity below.

The practical tasks of urban public policy and planning, then, can be typified as being directed to collective action problems in regard to the mobilization of resources, the consolidation of latent benefits, and the coordination of urban life in general, but always with the qualification that they are infused in various ways by the logic of urban space and by the logic of capitalism at large. Right from the beginnings of industrial urbanism collective action has been necessary to deal with the technical breakdowns in large cities stemming from their dynamics of growth and internal readjustment, like congestion, pollution, public health crises, land use conflicts, neighborhood decay, etc. (Benevolo 1971). These breakdowns are essentially diseconomies of urbanization that in the absence of at least partial remedial action would rapidly impose barriers to further urban expansion and hence accumulation in general. But in addition to clearing away physical impediments to growth and social viability, urban public policy is also frequently directed to the search for strategic outcomes that would simply fail to materialize, or would appear only in stunted form if competitive market order alone prevailed. Here, a plethora of possibilities might be enumerated, ranging from the implementation of infant industry programs on the economic side to communal development projects on the social side. There can be no doubt that the possibility of governance failures is always present in initiatives like these, and careful pre-policy analysis of the relevant empirical situation is hence a necessary, but certainly not sufficient, condition for success.

The tasks of public regulation are made yet more urgent by the structures of cumulative causation that underlie urban growth patterns and by the relatively slow convertibility of urban land uses. These dynamic properties of cities mean that they are endemically subject to path-dependent trajectories of evolution, which means in turn that some further degree of policy oversight is desirable in the effort to guard against negative lock-in effects over time. The more general point can be advanced to the effect that a purely market-driven *optimum optimorum* of urban outcomes is impossible; the best that can be achieved under market arrangements alone is some local equilibrium of a few fast-acting variables, leaving the rest of the urban system locked into market failure, systematic under-performance, and recursive inertia over time. In these circumstances, urban growth and development are inevitably susceptible to severe curtailment in the absence of adjunct frameworks of policymaking and planning. At this stage, it is well to recall one of the essential messages conveyed earlier in the discussion, namely, that these frameworks never operate on a purely technocratic basis (even though they almost always have strong technocratic elements), for policymakers and planners are continually subject to a tug-of-war between many different priorities resulting from both the vertical and horizontal stratification of urban society and the consequent contestation that occurs between opposing social and spatial constituencies over the direct and indirect distributional effects of public action. To be sure the intensity of this contestation varies widely over space and time.

The policy and planning problems posed by the internal crises of modern cities have both recurrent and conjunctural rhythms just as they have both local and national dimensions, and these different time–space registers leave distinctive marks on policymaking and policy-implementation arrangements. On the one hand, generic types of management and control measures are required to deal with the chronic problems (such as congestion, disorderly land use, or neighborhood decline) of the urban environment. On the other hand, many urban policy and planning initiatives are more episodic in character in that they are specific to a certain historical moment and the particular ways in which its associated social and political stresses intersect with the urban process. A striking illustration of this point is provided by the keynesian welfare-statist policy apparatus that was put into effect by the government of the United States in the decades following World War II. Keynesian welfare-statism was in the first instance a *national* policy designed to alleviate the malfunctions of fordist mass-production society as a whole. But it was in significant degree translated into practical outcomes by means of explicitly *urban* projects not only because so much of the mass-production system itself was deeply embedded in the large cities of the Manufacturing Belt but also because the increasingly ill-adapted infrastructures and inadequate housing arrangements of the same cities were themselves a significant part of the problem (Brenner 2004). Thus, intra-urban highway construction programs, urban renewal, and the planned expansion of the suburban housing stock, among other planning initiatives undertaken over the 1950s and 1960s, functioned as hybrid expressions of national policy imperatives and localized instruments of urban regeneration.

From all of the above, it follows that we can best understand the formulation and implementation of public policy and planning measures in the modern city in terms of two main interrelated lines of force. First, they function as corrective responses to ascertainable forms of urban disorder brought on by the very logic and dynamics of urbanization itself. Second and concomitantly, they are instruments for proactive intervention, as represented, for example, by the establishment of coordinating mechanisms to secure economic gains that would fail to emerge in the absence of collective action. In any case, they respond to problems and opportunities that occur in the urban system and that imperil, in one way or another, overall processes of economic accumulation and social reproduction. By the same token, their range of operation is circumscribed by and channeled through a complex network of political norms, expectations, and pressures in society as a whole. It is precisely the absence of disciplined attention to these indicative moments that explains why so many of the more prophetic statements about the role and functions of urban policy and planning must be taken with a grain of salt. But once this judgment has been advanced, where, we might ask, does it leave us in terms of normative recommendations and the possibility of a progressive politics of the urban today?

Urban Dynamics and Policy Dilemmas Today: Some Key Issues

The core of the urbanization process in modern society flows from the basic (but certainly not all-encompassing) phenomena of production and work as structured at the macrolevel by prevailing capitalist social and property relations. At an earlier moment of history when fordist mass production and its large-scale growth-pole industries dominated the economic order of large American cities, urban policy was deeply interwoven with the national keynesian welfare-statist measures that so successfully underpinned this particular regime. Over the last few decades various transformations of this previous order of things have occurred. Modes of economic production in the more economically advanced countries have now shifted radically away from a dominantly fordist pattern—a circumstance that is also associated with increased general levels of economic competition as well as uncertainty and risk. Globalization also continues to expand apace, leading to many new threats and positive possibilities for urban production systems and steadily destabilizing the boundaries of the national economy as a frame of reference for economic organization and policymaking. And finally, in today's predominantly neoliberal policy environment, national governments are increasingly unable or unwilling to provide policy services to all the sectional and regional interests that find themselves under stress as a result of these changing economic and social winds. Many cities are thus experiencing major internal functional transformations, and are under unprecedented pressures to take the initiative in building local institutions and agencies to secure their own future prosperity.

The leading edges of economic growth and innovation today coincide increasingly with sectors in which intellectual and human capital, complemented by digital technologies, is becoming the key ingredient of production processes and a prime requisite of competitive success. These sectors represent the avant-garde of the cognitive-cultural economy, and while they may be found in cities of many different sizes, they occur above all in major metropolitan regions, where they often form strikingly dense and specialized clusters in the wider tissue of urban space. The centripetal pull of these clusters is much reinforced by the persistently transactions-intensive nature of the economic activities that they harbor, and by the high levels of competition and uncertainty that characterize so much of the new economy. As a consequence, and in view of the growth and employment capacities of the new cognitive-cultural economy, it is scarcely surprising to note that urban policymakers have recently seized enthusiastically on its promise as an instrument of local economic development. One of the first major segments of the new economy to be seen in this light was high-technology industry in the 1980s, and much was made of its potentials for stimulating regional expansion, even if, in many cases, the claims about its miraculous powers of economic revitalization were greatly exaggerated (cf. Miller and Côte 1987). In more recent years policymakers all over the world have also been turning their attention to creative or cultural-products industries as promising avenues to urban prosperity.[1]

Cognitive-cultural sectors of all varieties are clearly now rising to the top of the agendas of local economic development agencies, not only because they offer skilled, high-wage jobs but also because they are in numerous (but not all) instances both environmentally friendly and fountainheads of community-wide prestige. Not least of their attractions to urban policymakers is their partiality for locations in dense metropolitan areas and their job-creating capacities at a time when so many other kinds of economic activity are fleeing from these areas to more peripheral parts of the world. As a consequence, various experiments are now going forward in numerous cities in the effort to work out effective policy measures for sustaining local competitive advantages in these and allied sectors. These experiments entail, in particular, more or less sophisticated efforts to reinforce collective assets in such domains of local economic activity as value-added networks, the employment system, and the regional innovation processes, to mention only some of the most obvious (see, e.g., Bianchi 1992; Cooke and Morgan 1998; OECD 2001; Storper 1997). A number of cities have also sought to advance their ambitions in

---

[1] Representative policy statements about the local economic development possibilities of cultural-products industries can be found, e.g., in British Department of Culture Media and Sport (2001), Hong Kong Central Policy Unit (2003), IAURIF (2006), and STADTart (2000).

this matter by means of lavish public spending on large-scale arts and leisure projects, and in this manner not only to promote a new cognitive-cultural economy but also to enhance their function as key centers of global cultural influence (and, as a corollary, their role as magnets for large-scale investment and the in-migration of elite workers). Many cities, especially in North America, Europe, and Asia, are now moving rapidly in this direction. To cite just one example, several of the Urban Regeneration Companies that have been promoted in recent years by the British Labour government and established by local partners have put a high degree of reliance on the economic development capacities of cognitive-cultural sectors. Even many cities in the erstwhile world periphery—Beijing, Shanghai, Hong Kong, Seoul, Rio de Janeiro, and Buenos Aires come readily to mind—are gearing up for similar major initiatives. Singapore, once a colonial entrepôt center and then a major depot of electronic assembly operations under the aegis of US, European, and Japanese multinationals, now brands itself the "global city of the arts" (Chang 2000).

The advent of the cognitive-cultural economy has therefore been accompanied by many bright new prospects for cities. At the same time, this more positive aspect of the current conjuncture is complemented by a very much more somber set of outcomes, for alongside the well-paid jobs that are appearing as the new economy expands, large numbers of low-wage jobs are also being generated. The latter jobs are to be found especially in the informal and underground economy of the city. Modern cities have always been characterized by a glaring divide between upper and lower income groups, but the divide has tended to widen significantly in large cities over the last couple of decades as the new economy has moved forward (Fainstein 2001; Hamnett and Cross 1998). The social tensions that crystallize around this phenomenon are exacerbated by the fact that so much of the workforce in the low-wage segment of the contemporary urban economy is composed of immigrants from less developed countries. A large section of this workforce constitutes a socially marginalized and politically disenfranchised mass of individuals whose position on the fringes of urban society is further underlined by the high risks of unemployment and underemployment that they face. The problem is even more acute in the case of African-Americans in US cities, for many of these individuals appear to face potent structural barriers to any kind of employment whatever (see Chapter 6). Periodically, the stresses inherent in this situation break out in explosions of rioting and unrest. Emergency policing and stopgap measures may well bring spontaneous disturbances of this sort under short-term control, but the situation from which they stem represents nevertheless a simmering, long-term, and multifaceted problem that urgently calls for deep-seated remediation. This problem of the underclass in modern cities raises fundamental issues that lie far beyond any immediate concerns about technocratic procedures of social restraint. Above and beyond the need for improved employment conditions and opportunities for low-wage workers, these issues go directly to a number of pressing concerns at the heart of contemporary urban society, in particular, political representation, distributional equity, and the democratization of urban space. I shall have much to say about these matters in Chapter 6. For the present, it is worth noting that resolution of these concerns is not simply an issue of social fair play, important as that issue may be in its own right. In addition, the full potentials of urban economic development in the era of cognitive-cultural capitalism are apt to be severely constrained so long as there are large segments of the citizenry condemned to second-class status.

One further overarching policy challenge posed by the continued rapid growth of cities in the context of globalization revolves around a series of new governance imperatives in the interests of competitiveness and social order (MacLeod 2001). I have already mentioned the balkanization of urban administrative activities and the reflection of this state of affairs in metropolitan-wide patchworks of independent municipalities. Balkanization of this sort has always presented managerial challenges in cities, if only in the narrow sense that unregulated inter-municipal spillover effects are typically widespread in any given metropolitan area, but it has assumed expanded significance under conditions of increasing global competition. The complexity of this situation is compounded by two additional developments. For one thing, national restraints on urban growth and development in the advanced economies have relaxed considerably by comparison with conditions in the 1960s and 1970s when territorial equalization was very much on the political agenda. For another, the application of subsidiarity principles is leading to increasing devolution of much social and economic management to the urban level, with the consequence that city administrations are more than ever before confronted with enormous burdens in regard to the formulation and implementation of policy. Large cities everywhere are struggling to face up to these circumstances, above all, perhaps, in regard to the pressing need for institution-building in support of localized competitive advantages. As Jonas and Pincetl (2006) have intimated, however, the growing mismatch that is observable between the internal social and economic organization of the metropolis on the one hand, and its fragmented political geography on the other, puts shackles on the possibilities for decisive and concerted action.

More effective policymaking and institutional arrangements at the intra-metropolitan level are, of course, essential given the interdependencies that run through the internal organism of the city. They are more particularly imperative in a globalizing world where cities are open to the gales of international competition, and where so much of their ability to react to and rise above these gales depends on an enhanced capacity both to manage their existing economic assets and to strike out with new initiatives for positive action within their own jurisdictions. Nowhere is this need more pressing than in those cities that now play an increasing role as "national champions" (Jessop 2004) and as motors of the new global economy.

3

Production and Work in the American Metropolis

Introduction

In much of the foregoing discussion I have made a special point of claiming that a new cognitive-cultural economy seems to be moving to the fore in the advanced capitalist countries. This new economy is bursting forth with special vigor in the major cities of the United States. As it does so, more standardized production activities (e.g. many types of manufacturing) are declining in both absolute and relative significance in the largest metropolitan areas. The decline is accentuated by the transfer of a large proportion of these more standardized activities to low-cost locations offshore. At the same time, smaller metropolitan areas in the United States continue to exhibit comparatively high levels of specialization in manufacturing. The net consequence of these developments is that there are rather marked shifts in the economic character of cities as we move from the largest to the smallest centers in the urban hierarchy. We now turn to an empirical investigation of these initial claims in order to ground the discussion in a firmer sense of the economic logic of urbanization and as an entry point into deeper examination of some of the radically new urban forms and functions coming into being in America's most dynamic cities.

The argument moves ahead, then, on the basis of a statistical analysis of the economic structure of American metropolitan areas, with particular, but not exclusive focus on the geography of cognitive-cultural production. We shall also be paying attention to the ways in which this form of production appears to be displacing an older division of labor in the very largest cities, and to be ushering in new modalities of urban social stratification. This, by the way, is not to say that the division of labor has been abolished in the new economy, only that it has tended in many cases to move away from the minute parcelization of tasks characteristic of the smithian version and to allow for more worker discretion in labor processes and more variable interactions between employees, as represented most dramatically perhaps by project-oriented work. On the descriptive side, I offer a broad panorama of the sectors and occupations (or, rather, segments of sectors and occupations) that are to be found in specific locational niches within the overall hierarchy of metropolitan areas. On the analytical side, I identify a number of the technological, organizational, and job-related characteristics that govern the locational sorting out of economic activities within the metropolitan hierarchy, with special reference to cognitive-cultural versus more routinized and standardized forms of work. I argue in general that there is a broad bias toward the former in the economies of large metropolitan areas, while in small metropolitan areas the bias is toward the latter. My main hypothesis, in a nutshell, is that there is a definite gradation in frameworks of competitive advantage and hence in the nature of production and work across the urban hierarchy, from the largest centers (where labor-intensive, innovative, and customized production is relatively common) to the smallest (where more capital-intensive and repetitive types of economic activity prevail).

The general form of this hypothesis is by no means new (cf. Blackley and Greytak 1986; Scott 1982) but the ready availability of large relevant data-sets now makes it possible to lay out the evidence in a very much more overarching and systematic manner than hitherto.[1] In the attempt to operationalize the investigation, I deploy two main bodies of data on US cities, one from the *Economic Census* of 2002 and the other from the *Decennial Census* of 2000, together with various other pieces of empirical information drawn from official sources. These data-sets, taken in combination with one another, provide a reasonably compelling and mutually confirming body of evidence in regard to the broad hypothesis that is under scrutiny here.

---

[1] Early intimations of this hypothesis can also be found, e.g., in Duncan et al. (1960) and Florence (1955).

Growth and Transformation of the Metropolitan Economy

Even as the fordist mass-production system was entering its climactic period of crisis in the late 1970s the seeds of an unprecedented urban and regional resurgence were being planted in many different parts of the world. Some of the earliest manifestations of this resurgence were detected in the rise of small-scale neo-artisanal industries in the cities of northeast and central Italy (Becattini 1987; Brusco 1982); other instances were documented by researchers focusing on technology-intensive production in selected areas of the United States and Western Europe (Breheny and McQuaid 1987; Markusen, Hall, and Glasmeier 1986; Scott 1986); and yet other symptoms of this trend were observed by scholars writing on the expansion of business and financial services in major urban centers (Daniels 1979; Noyelle and Stanback 1984). Subsequently, a number of researchers also began to make note of the prominent role played by the infor-mation economy and cultural-products industries in the resurgence of metropolitan areas in the advanced capitalist societies (Drennan 2002; Hutton 2004; Molotch 1996; Power 2002; Pratt 1997; Scott 1996a; Storper and Christopherson 1987). Each and every one of these different forms of economic activity depends in deeply signif-icant ways upon the cognitive and cultural capacities of the labor force, be it in the domains of scientific and technological knowledge, research capacity, industrial design, business and financial acumen, sensibility to the idiosyncrasies of others, inventiveness, story-telling ability, or even simple adaptability to a generally volatile and unpre-dictable work environment. Moreover, as these different activities have come to the fore in the new American economy, so have they settled above all in large metropolitan areas.

An initial evaluation of the latter point can be accomplished by consideration of two sets of location quotients for selected sectors (Table 3.1) and occupations (Table 3.2) in metropolitan areas in the United States today.[2] Table 3.1 (and all subsequent tables showing sectoral data) is based on information taken from the *Economic Census*

[2] The location quotient is defined as follows. Let E_{ij} represent employment in sector or occupation i in some spatial unit j and let E_{i*} represent employment in sector or occupation i in the country as a whole. The location quotient LQ_{ij} is then equal to $(E_{ij}/\Sigma_i E_{ij})/(E_{i*}/\Sigma_i E_{i*})$. A value of LQ_{ij} greater than 1 indicates that sector or occupation i is overrepresented in spatial unit j relative to its incidence in the national economy as a whole; a value of LQ_{ij} less than one indicates that the sector or occupation is underrepresented relative to its incidence in the national economy.

Table 3.1. Location quotients for selected 2-digit industries by MSA population size category, 2002

NAICS sector	Total US employment (thousands)	Location quotients by metropolitan size category					
		>5M	1M–5M	500Th–1M	250Th–500Th	<250Th	Micropolitan areas
31–33 Manufacturing	14,700	0.83	0.85	0.96	1.16	1.18	1.61
42 Wholesale trade	5,878	1.21	1.10	1.01	0.89	0.81	0.82
51 Information	3,736	1.38	1.12	0.91	0.80	0.77	0.56
52 Finance and insurance	6,579	1.31	1.12	1.00	0.89	0.80	0.56
54 Professional, scientific, and technical services	7,243	1.44	0.97	0.82	0.81	0.64	0.53
62 Health care and social assistance	15,052	0.95	0.94	1.08	1.12	1.27	1.17
71 Arts, entertainment, and recreation	1,849	1.07	1.13	0.89	0.89	0.92	1.02
Number of metropolitan/micropolitan areas	—	9	37	29	49	109	526

Note: Differences between Table 3.1 and Table 3.2 in the number of metropolitan areas representing each size category can be accounted for by discrepancies in the sources used.

Source: US Bureau of the Census, *Economic Census*, 2002, accessed via http://factfinder.census.gov/home/saff/main.html?_lang=en.

Table 3.2. Location quotients for selected occupations by CMSA and MSA population size category, 2000

Occupations: Census codes and description	Total US employment (000)	Location quotients by CMSA/MSA size category				
		>5M	1M–5M	500Th–1M	250Th–500Th	<250Th
001–099 Management, business and financial operations occupations	17,448	1.15	1.06	0.92	0.88	0.83
050–099 Business and financial operations occupations	5,559	1.23	1.13	0.94	0.87	0.77
100–359 Professional and related occupations	26,199	1.13	1.01	1.00	0.96	0.98
100–129 Computer and mathematical operations	3,168	1.44	1.18	0.85	0.74	0.60
130–153 Architects, surveyors, cartographers, and engineers	1,927	1.22	1.13	1.04	0.95	0.78
154–156 Drafters, engineering technicians, and mapping technicians	733	0.90	1.08	1.09	1.11	1.06
210–219 Legal occupations	1,412	1.41	1.07	0.93	0.81	0.66
260–299 Arts, design, entertainment, sports, and media occupations	2,484	1.36	1.01	0.84	0.84	0.82
300–359 Healthcare practitioners and technical occupations	5,980	0.98	1.00	1.09	1.07	1.14
360–469 Service occupations	19,277	0.95	0.97	1.01	1.05	1.08
360–369 Healthcare support occupations	2,593	0.94	0.88	1.03	1.02	1.08
430–469 Personal care and service occupations	3,628	1.00	0.99	0.94	1.03	1.05
470–599 Sales and office occupations	34,621	1.02	1.06	1.03	1.00	0.99
470–499 Sales and related occupations	14,592	1.00	1.06	1.02	1.03	1.02
500–599 Office and administrative support occupations	20,028	1.04	1.05	1.03	0.98	0.96
770–979 Production, transportation and material moving occupations	18,968	0.83	0.88	1.00	1.06	1.08
770–899 Production occupations	11,008	0.82	0.85	0.99	1.07	1.09
900–979 Transportation and material moving occupations	7,960	0.86	0.93	1.02	1.04	1.07
Total number of CMSAs/MSAs	—	9	41	33	65	133

Note: Differences between Table 3.1 and Table 3.2 in the number of metropolitan areas representing each size category can be accounted for by discrepancies in the sources used.

Source: US Bureau of the Census, *Decennial Census*, 2000, accessed via http://factfinder.census.gov/servlet/DownloadDatasetServlet?_lang=en.

of 2002; Table 3.2 (and all subsequent tables showing occupational data) is based on information taken from the *Decennial Census* of 2000. In the former case, much of the data must be estimated on the basis of the information-suppression codes used by the Bureau of the Census in order to ensure confidentiality in cases where the original reporters might be individually identifiable. There is thus a derivative margin of error in all of the sectoral data under examination here and in the location quotients calculated from them. Data from both of these sources are organized by metropolitan size categories, of which I have identified five, that is in terms of population: (*a*) 5,000,000 and above, (*b*) 1,000,000–5,000,000, (*c*) 500,000–1,000,000, (*d*) 250,000–500,000, and (*e*) 250,000 and below. In addition, data from the *Economic Census* make it possible to identify a sixth category of urban centers, namely micropolitan areas.

Table 3.1 lays out location quotients for seven major 2-digit sectors and the six metropolitan/micropolitan categories identified above.[3] The sectors selected here were chosen on the basis of both their essential importance in the modern US economy and their presumed interest as core indicators of the changing structure of metropolitan production systems. Collectively they represent 52.2 percent of total US employment. Sectors not selected for admission into Table 3.1 were omitted either because they are resource-based industries or because they represent nonbasic activities that are relatively evenly spread out across the entire space-economy.[4] The 2-digit sectors shown in Table 3.1 represent a wide array of industrial types, ranging from manufacturing to finance and insurance, and from wholesale trade to arts, entertainment, and recreation. Two main points now

[3] Note that these size categories are meant to represent generic types of metropolitan area, and cannot be taken to stand in for any one case. Nevertheless, it is useful to inquire as to what degree they characterize individual cases and/or to what degree they may be biased. In this regard, two simple diagnostic tests were carried out relative to the nine metropolitan areas that make up the largest size category. These two tests consist in a comparison of employment distributions in each metropolitan area of all 2-digit and 3-digit sectors with the equivalent distributions for all nine areas taken collectively (where all data are taken from the *Economic Census* of 2002). A Kolmogorov–Smirnoff test suggests with a 99% level of confidence that the individual distributions are identical to the relevant collective distribution. By contrast, we can expect that as we move down the metropolitan hierarchy, certain individual cities will tend to diverge significantly from the collective representation. I refer briefly to the latter point later in the text.

[4] The specific 2-digit sectors omitted from the analysis are 21 (Mining), 22 (Utilities), 44–45 (Retail trade), 48–49 (Transportation and warehousing), 53 (Real estate and rental and leasing), 55 (Management of companies and enterprises), 56 (Administrative and support and waste management and remediation), 61 (Educational services), 72 (Accommodation and food services), and 81 (Other services, except public administration).

need to be made. First, manufacturing together with healthcare and social assistance are the only sectors shown in Table 3.1 for which the computed location quotients vary inversely with metropolitan size category. This relationship is very strongly evident in the case of manufacturing, but is more subdued in the case of healthcare and social assistance. Second, the remaining five sectors (wholesale trade; information; finance and insurance; professional, scientific, and technical services; and arts, entertainment, and recreation) are all characterized by location quotients that change positively in relation to metropolitan size class. In view of the general character of these five sectors, the results here appear to be consistent with our initial hypothesis that less standardized forms of production are apt to locate in large metropolitan areas, whereas more standardized forms are apt to be found in small metropolitan areas. They are also consistent with our sub-hypothesis to the effect that the most insistently cognitive-cultural sectors will tend to gravitate toward large metropolitan areas. That said, the results also contain a number of special cases. The healthcare and social assistance sector, as noted, is unusually well represented in small cities, in part, no doubt, because of a relative buildup of publicly supported social services in these areas. Conversely, wholesale trade is concentrated in large metropolitan areas presumably because of its need for nodal locations from which distribution can be effectively carried out. At the same time, and contrary to expectations, the arts, entertainment, and recreation sector is not as strongly differentiated across the urban hierarchy as the other main cognitive-cultural sectors under examination here, but this peculiarity turns out on closer inspection to be a function of the very high level of aggregation that characterizes this sector in which relatively recondite activities (e.g. theater and dance companies) are combined with other activities that are much more down to earth (e.g. amusement arcades and bowling centers).

The information laid out in Table 3.2 now turns the spotlight on the occupational dimensions of these issues. In this table, location quotients are shown for major census occupational groups and sub-groups cross-tabulated by metropolitan size categories. The evidence arrayed in Table 3.2 runs more or less parallel to the information presented in Table 3.1. Occupations that entail significant amounts of routine blue-collar work (i.e. production, transportation and material moving occupations) are clearly most prominent in small metropolitan areas. By contrast, occupations that have conspicuous cognitive-cultural attributes (e.g. management etc. occupations or professional and related occupations) are much more important in large metropolitan areas. Lower-level service occupations such as healthcare support and personal care, together with sales and office occupations, are rather evenly represented across all metropolitan size categories, though the gradient of the relationship varies mildly in each case in a manner that seems quite reasonable (e.g. negative for healthcare support; positive for office and administrative support). It is of interest to note that the location quotients for drafters, engineering technicians, and mapping technicians are negatively correlated with metropolitan size category, even though these occupations are classified under the broader heading of professional and related occupations. Drafting etc. occupations have been especially susceptible to displacement by computerization over the last couple of decades, and have declined sharply in numbers of late years, especially in large metropolitan areas (cf. Skinner 2004). Again, then, and even at this crude level of analysis, the basic hypothesis laid out at the beginning of this chapter appears to stand up reasonably well to rough and ready empirical scrutiny in that large metropolitan areas—with certain plausible exceptions—are much more likely than small to evince an occupational profile marked by types of employment that demand high levels of discretionary cognitive and cultural performance on the part of workers.

A further broad piece of evidence to this effect can be adduced by consideration of educational levels as a function of metropolitan population. For 281 CMSAs and MSAs, the correlation coefficient for an equation linking the log-odds of the proportion of the population holding a bachelor's degree and the logarithm of population is a very significant 0.30, and for the proportion of the population holding a professional degree it is 0.38.[5] In other words, large metropolitan areas are significantly more endowed than small with pools of highly educated labor. In addition, Herfindahl indexes (based on numbers of workers employed) were computed across all 2-digit sectors for each of the 918 individual metropolitan and micropolitan statistical areas

---

[5] A simple log-odds regression equation can be written as $\ln[p/(1 - p)] = a + bx$ where p is a probability or proportion and x is an independent variable. After suitable manipulation, this equation becomes $p = \exp(a + bx)/[1 + \exp(a + bx)]$ or $1/[1 + \exp(-a - bx)]$. Thus, the computed values of a and b are the parameters of a logistic equation for p with the essential property that $0 \geq p \leq 1$.

mentioned in the *Economic Census* of 2002.[6] These indexes represent measures of diversity (when they are low) and specialization (when they are high). The computed indexes were then correlated with metropolitan population values, giving a correlation coefficient of –0.24, which is significant at well beyond the 0.01 level. This finding tells us that economic diversity (which is an element of competitive advantage) declines with size of metropolitan area. Herfindahl indexes were also computed across 31 major occupational groups (excluding farmers and farm managers) for the 281 CMSAs and MSAs that figure in the *Decennial Census* of 2000, and correlated with population values. On this occasion, however, a statistically insignificant correlation coefficient of –0.08 was the result. The evident deduction here is that major occupational groups are too broadly defined to pick up significant variations at this geographic scale. I should add that neither unemployment rates nor sex ratios in the labor force show much variation across different size categories of metropolitan areas, though sex ratios do, of course, vary greatly by sector and occupation.

Manufacturing Activities in the Contemporary Metropolis

We now turn our attention to a very much more detailed empirical evaluation of these matters. In this section, the focal point of analysis is concerned with the manufacturing side of the economy and with the functional–structural features that give rise to differentials in the locational incidence of manufacturing sectors across the urban hierarchy. In the subsequent section, we will examine a more general body of data based on the broad occupational characteristics of metropolitan areas.

We have already noted in general terms that the relative significance of manufacturing employment in any metropolitan area tends to increase as the size of metropolitan area decreases. This proposition is reinforced by consideration of Table 3.3, where changes in manufacturing employment by metropolitan size category between 1997 and 2002 are displayed. Manufacturing employment over this period declined across all metropolitan size categories, but most

---

[6] The Herfindahl index for any sector i, is defined as $H_i = \Sigma_j p_{ij}^2$, where p_{ij} designates the proportion of total employment in sector i that is located in metropolitan area j.

Table 3.3. Change in total manufacturing employment by metropolitan population size category, 1997–2002

MSA size category	Number of metropolitan areas[a]	Employment (000)		Percentage change
		1997	2002	
More than 5 million	9	4,000	2,837	−29.1
1 million–5 million	26	2,929	2,406	−17.8
500 thousand–1 million	30	1,164	958	−17.7
250 thousand–500 thousand	43	879	751	−14.6
50 thousand–250 thousand	114	1,077	1,034	−4.1
United States	—	16,888	14,703	−12.9

[a] NB: The number of metropolitan areas for which data are given differs widely from the 1997 to the 2002 *Economic Census*; only metropolitan areas that appear in both counts are used in the analysis here.

Source: US Bureau of the Census, *Economic Census*, 1997 and 2002, accessed via http://factfinder.census.gov/home/saff/main.html?_lang=en.

sharply in the largest and least sharply in the smallest. If this trend continues, the obvious outcome is that the inverse relationship between manufacturing location quotients and metropolitan size categories as observed in Table 3.1 will become yet more acute in the future.

Our immediate task is to explore a number of the minutiae underlying this trend. In pursuit of this goal, I seek to model the locational variation of different manufacturing sectors as a function of a series of basic technological and organizational variables. I accordingly present two parallel sets of regression equations that describe locational variations of 6-digit NAICS manufacturing sectors within the metropolitan hierarchy. The dependent variable in these equations is expressed as the incidence of 6-digit sectors in larger metropolitan areas relative to their incidence in smaller metropolitan areas. Thus, in one set of equations the dependent variable is p_{i1}/p_{i2} (or $p_{i1}/(1 - p_{i1})$), that is the proportion (p_{i1}) of *employment* in metropolitan areas with a population of over 5 million divided by the proportion (p_{i2}), of employment in all other metropolitan areas, where the subscript i runs over all twenty-one 3-digit sectors from NAICS 311 to NAICS 339; in the other set, the dependent variable is p'_{i1}/p'_{i2} (or $p'_{i1}/(1 - p'_{i1})$), that is the proportion (p'_{i1}) of total *establishments* in metropolitan areas with a population of over 5 million divided by the proportion (p'_{i2}) of establishments in all others. The independent variables used in the analysis are identified in Tables 3.4 and 3.5 and

Table 3.4. Log-odds regressions of the of proportion employment in large metropolitan areas (over 5 million people) relative to the proportion of employment in all other metropolitan areas, for 6-digit manufacturing sectors, 2002[a]

Regression coefficients	Regression coefficients	
	Model 1	Model 2
Capital–labor ratio	−0.217	−0.265*
Average size of establishments	−0.514**	−0.386**
Materials intensity	−0.196	−0.146
Machinery rentals per production worker	0.220*	0.304**
Inventory turns	0.289**	0.315**
Ratio of contracts to shipments	−0.047	0.101*
Percent production workers	−1.601**	−1.416**
Hourly wage	−0.379	−0.809*
Average annual salary	1.217**	0.964*
Fixed industry effects?	Yes	No
Constant	−3.538	−0.612
R^2	0.426	0.280
Adjusted R^2	0.328	0.264
N	404	404

[a] Observations are 6-digit manufacturing sectors; all independent variables have been transformed to natural logarithms; independent variables are defined in the Appendix.
*Significant at 0.05 level; **significant at 0.01 level.

are further defined in the Appendix. These variables refer to three main dimensions of sectoral variation in manufacturing, namely, (*a*) investments and scale (capital–labor ratio, average size of establishment, and materials intensity), (*b*) levels of stability/instability and externalization in production (machinery rentals, inventory turns, and the ratio of contracts to shipments), and (*c*) employment and remuneration (production workers as a percentage of total employment, hourly wages, and average annual salary). The specific relevance of these variables to urban variations in the manufacturing economy is traced out below. In addition, a fourth set of independent variables is deployed in the analysis, and these consist simply of {0,1} fixed industry effects, or dummies, one for each of the 3-digit manufacturing sectors in the NAICS nomenclature. Data for a total of four hundred and four 6-digit sectors were extracted from the *Economic Census* of 2002. Given the definition of the dependent variables, the regressions are computed in the form of log-odds equations, as laid out in Tables 3.4 and 3.5.

Table 3.5. Log-odds regressions of the proportion of establishments in large metropolitan areas (over 5 million people) relative to the proportion of establishments in all other metropolitan areas, for 6-digit manufacturing sectors, 2002[a]

Independent variables	Regression coefficients	
	Model 1	Model 2
Capital–labor ratio	0.181	−0.087
Average size of establishments	−0.555**	−0.312**
Materials intensity	−0.254*	−0.141
Machinery rentals per production worker	0.054	0.106
Inventory turns	0.205*	0.211*
Ratio of contracts to shipments	−0.062	0.098*
Percent production workers	−0.928**	−1.188**
Hourly wage	−1.359**	−1.563**
Average annual salary	1.173**	1.068*
Fixed industry effects?	Yes	No
Constant	0.806	1.301
R^2	0.406	0.213
Adjusted R^2	0.360	0.194
N	404	404

[a] Observations are 6-digit manufacturing sectors; all independent variables have been transformed to natural logarithms; independent variables are defined in the Appendix.
* Significant at 0.05 level; **significant at 0.01 level.

Scrutiny of Tables 3.4 and 3.5 reveals a remarkably robust and theoretically meaningful set of results. In all cases values of the adjusted R^2 are very respectable in magnitude and extremely significant in statistical terms, and most of the regression coefficients presented in the tables are also highly significant. Both tables contain two principal equations, representing regression results with and without the full set of 21 fixed industry effects. It is particularly gratifying to note that the regression coefficients attached to the main independent variables are extremely stable both in magnitude and sign irrespective of the presence or absence of fixed industry effects. The substantive results of the regression analysis can best be conveyed in four main steps as follows.

First, the equations in Tables 3.4 and 3.5 indicate that the attraction of any given 6-digit manufacturing sector to locations in large as opposed to small metropolitan areas is strongly but inversely related to levels of investment and scale, which in turn are roughly symptomatic of degrees of overall flexibility versus routinization in any given sector. The lower the capital–labor ratio, the average size of establishments, and the materials intensity of production, the greater is the likelihood that the sector will be found in large metropolitan areas; higher values of these three variables increase the likelihood that it will be found in small metropolitan areas. Average size of establishments is the most effective predictor in this sequence of variables, and this accords well with much of the case-study literature in which the spatial concentration of small vertically disintegrated plants in large cities is frequently stressed (Scott 1982).

Second, the relative incidence of any 6-digit sector in large metropolitan areas is positively related to machinery rentals per production worker, inventory turns, and (somewhat more ambiguously) the ratio of contracts to shipments. The interpretation of this finding is that sectors marked by short time horizons (as expressed in the rental rather than ownership of basic equipment) combined with high levels of instability and a proclivity to outsourcing will locate with a high probability in large metropolitan areas, whereas sectors with the opposite characteristics will tend to locate in smaller metropolitan areas.

Third, the odds that units of production in any 6-digit manufacturing sector will locate in larger rather than smaller metropolitan areas are greatly increased if the percentage of production workers in total employment is low, if hourly wages of production workers are low, and if average annual salaries of nonproduction workers are high. These three variables indicate that manufacturing employment in the large metropolis tends to be biased toward relatively skilled nonproduction workers, who are paid high salaries in comparison to their counterparts in small metropolitan areas, while manufacturing employment in the small metropolis is biased toward production workers, who are paid high wages in comparison to their counterparts in large metropolitan areas. The finding that manufacturing sectors in the large metropolis are associated with relatively low average wages for production workers is doubtless explicable in the light of two main factors, one linked to the buildup of politically marginalized immigrant workers in the labor force of major urban centers in America and the other linked to the high levels of competition and relatively short-term employment contracts that evidently characterize lower-tier labor markets in the same centers (Borjas 2003; Jayet 1983; Ward 2005).

53

Fourth, the 21 fixed industry effects obviously play an important role in the overall statistical analysis, suggesting that there is substantial idiosyncrasy in locational patterns from one 6-digit manufacturing sector to another. The role of three of the fixed effects merits special attention in the present context. Both of the dependent variables under analysis here are very significantly conditioned by the fixed effects for NAICS 315 (Apparel Manufacturing), NAICS 321 (Wood Product Manufacturing), and NAICS 312 (Beverage and Tobacco Product Manufacturing). In the case of the apparel industry the relationship is positive, a circumstance that can probably be ascribed to the notably deroutinized and destandarized mode of operation of much of this sector, especially in view of the fact that the more regimented side of the industry has been steadily moving offshore in recent years. Both wood product manufacturing and beverage and tobacco product manufacturing are to a significant degree resource-intensive industries, and the observed negative relationship in this instance almost certainly reflects the locational pull of sites where appropriate materials inputs are readily available.

Notwithstanding the heterogeneity of the data used in these regressions, the results are very satisfactory. While metropolitan areas in the United States may in general be losing manufacturing employment at a rapid rate, they nonetheless are subject to definite patterns of locational sorting and differentiation as a function of size (see also Pollard and Storper 1996; Rigby and Essletzbichler 2005). Large metropolitan areas, in particular, remain quite hospitable to labor-intensive segments of manufacturing industry, especially in sectors such as clothing, jewelry, tool and die production, electronic components, and so on, that is, sectors where establishments are on average small in size and where both internal and external production relations are subject to frequent shifts. In large metropolitan areas, many of these same sectors are also generally associated with a bifurcated labor force comprising relatively well-paid nonproduction and artisanal workers on the one side and low-wage production workers on the other, and this circumstance no doubt reflects their need for highly qualified labor inputs in such domains as technology, design, craft work, and commercialization, combined with a heavy reliance on immigrant workers to carry out ever-shifting successions of more mundane tasks. As we move down the urban hierarchy, manufacturing activities shift into more capital-intensive and routinized modes of operation in generally larger establishments. Simultaneously, the labor force is increasingly composed of production workers earning relatively high hourly wages while the salaries of nonproduction workers tend to decline compared with their homologues in larger centers, possibly as a function of their lower overall levels of qualification. One final point in this context is that value-added per worker was found to be higher in larger metropolitan areas, but played no significant role in the equations presented in Tables 3.4 and 3.5.

The overall hypothesis driving this analysis forward is thus quite strongly supported by these findings. Large metropolitan areas are indeed found to be foci of destandardized, labor-intensive production units caught up in flexible networks of interaction, in comparison with small metropolitan areas where more capital-intensive and routinized manufacturing activities tend to locate. Moreover, even though manufacturing as a whole lies toward the bottom end of the cognitive-cultural spectrum, the results laid out here suggest that even in this instance, a significant cognitive-cultural dimension (especially in technology-, design-, and fashion-intensive sectors) typifies the manufacturing economies of larger metropolitan areas.

The Occupational Structure of the Metropolitan Economy

We now push yet more deeply into the inquiry by means of a finely grained analysis of the occupational structure of American metropolitan areas. Investigation of detailed occupational categories is important and useful in its own right in the present context, but it has the further advantage that we can attach to these categories a series of diagnostic indexes that reveal much about the substantive cognitive and cultural content of different labor tasks. These indexes are drawn from the *Dictionary of Occupational Titles*, published by the US Department of Labor.

The *Dictionary of Occupational Titles* was last revised in 1991, and has now actually been superseded by the vastly more ambitious O*Net Program.[7] Nevertheless, the *Dictionary* offers a unique set of insights into the qualitative attributes of different occupations by reason of the variables that it provides to describe and measure the functional relationships of workers in the workplace to *data*, *people*, and *things*. Table 3.6 lays out the component elements of

---

[7] O*Net is accessible at http://online.onetcenter.org/.

Table 3.6. Ranking of occupational tasks in relation to data, people, and things

Data	People	Things
0 Synthesizing	0 Mentoring	0 Setting up
1 Coordinating	1 Negotiating	1 Precision working
2 Analyzing	2 Instructing	2 Operating–controlling
3 Compiling	3 Supervising	3 Driving–operating
4 Computing	4 Diverting	4 Manipulating
5 Copying	5 Persuading	5 Tending
6 Comparing	6 Speaking-signaling	6 Feeding–offbearing
	7 Serving	7 Handling
	8 Taking instructions	

Source: US Department of Labor, *Dictionary of Occupational Titles* (Fourth Edition) 1991, accessed at http://www.oalj.dol.gov/libdot.htm#appendices.

each of these variables along with their assigned numerical rankings[8] which—and this is important for the subsequent discussion—are defined so that low rankings designate more complex tasks and high rankings designate less. The data variable is differentiated along seven main axes running from synthesizing (ranked 0) and coordinating (ranked 1), which involve high-level thought processes and analytical capacities, to copying (5) and comparing (6), which involve only elementary forms of data manipulation. The people variable runs from mentoring (0) and negotiating (1), which call for detailed and subtle personal interactions, to serving (7) and taking instructions (8), which are very much less demanding in terms of interpersonal skills. The things variable runs from setting up (0) and precision working (1), which are dependent on both finely honed technical knowledge and skilled hand-eye coordination, to feeding-offbearing (6) and handling (7), which are relatively simple manual operations. Every occupation given in the *Dictionary* is then identified with three scores, one each for the data, people, and things variables.

Unfortunately, the occupational categories in the *Dictionary* are somewhat different from those used in the *Decennial Census* of 2000, though it is a relatively straightforward if laborious matter to match the two. Ambiguous cases in this matching exercise could usually but not always be resolved by reference to detailed

---

[8] These rankings are deployed in the official code for every occupation as laid out in the *Dictionary of Occupational Titles*. The fourth digit of the code identifies a specific kind of relation to data, the fifth to people, and the sixth to things.

occupational descriptions. To ease the task of matching, occupational titles for which national employment fell below 20,000 in the year 2000 were removed from consideration. Occupations related to (a) farming, fishing, and forestry, (b) protective services, (c) construction and extraction, (d) transportation and materials moving, and (e) the military were also discarded at the outset, either because they have little bearing on the dynamics of the metropolitan economy as such or because they are a systematically recurrent element of every metropolitan system. In the end, a total of 362 detailed occupational titles from the *Decennial Census* of 2000 were matched to titles given by the *Dictionary*, and hence were also coded by reference to their relation to data, people, and things. Because of the time lag between the last edition of the *Dictionary* and the data for the year 2000, actual work tasks within a number of occupations will probably have evolved somewhat in their relations to data, people, and things, and for this reason we need to proceed with a degree of caution in what follows.

It is important at this stage to inquire further into the specific meanings attached to the data, people, and things variables. An initial evaluation of this matter can be made by considering the statistical interplay between these variables for the 362 occupations retained for examination here. Table 3.7 shows the results of a rank correlation analysis of the three variables. This exercise reveals that the data and people variables have a significant degree of positive overlap with one another, as might be anticipated in view of their strong cognitive and cultural resonances. The things variable appears at the outset to be completely independent of the data variable and negatively related to the people variable. On closer examination, however, a conspicuous anomaly can be found in the things variable. It turns out

Table 3.7. Rank order correlations between occupational tasks[a]

Variables	Data	People	Things	Things (less category 7)
Data	1.00	—	—	0.68**
People	0.50**	1.00	—	0.20**
Things	0.01	−0.19**	1.00	—
Variable means	2.78	6.09	4.30	2.25
N	362	362	362	206

[a]Observations consist of the scores of each function on selected occupational categories.
* Significant at 0.05 level; **significant at 0.01 level.

Table 3.8. Correlations between occupational task ranks and specified levels of educational attainment[a]

Level of educational attainment	Data	People	Things	Things (less category 7)
Bachelor's degree	−0.61*	−0.32**	0.21**	−0.19**
Master's degree	−0.49*	−0.40**	0.22**	−0.29**
Professional degree	−0.17*	−0.43**	−0.11*	−0.14

[a]Observations consist of the percentage of labor force attaining the given educational levels in selected occupations.
*Significant at 0.05 level; **significant at 0.01 level.

that the *Dictionary of Occupational Titles* ranks many otherwise highly skilled jobs (journalism, for example or executive management) on the things variable at the level 7 (i.e. handling) which is no doubt correct insofar as it goes, but which skews the overall analysis in an unfortunate and unreasonable manner by aligning these occupations with unskilled manual workers. If we eliminate from consideration the 156 occupations that are coded at level 7 on the things variable, its relationship to the other two variables changes dramatically. The things variable now has a high positive correlation with the data variable and a somewhat lower positive correlation with the people variable (see Table 3.7). After adjustment, then, there is in the end a good deal of positive interaction between all three variables, though as will be made evident below, each of them offers a rather unique view of the operational features of individual occupations. Above all, these variables give us a rough but useful sense of the analytic, interactive, and physical dimensions of any given occupation. Table 3.8 reveals, in addition, that the three variables have a strong relationship to education, with low scores (representing nonroutine and more skilled work) being associated with high levels of education and high scores (representing routine and less skilled work) with lower levels. The things variable at first looks as if it runs counter to this remark, but after correction by again eliminating occupations which are scored 7, it behaves quite normally.

Table 3.9 and Figure 3.1 provide the main analytical results that follow from these preparatory remarks. At the outset, a vector of location quotients for all 362 detailed occupational categories was computed for each of the 5 metropolitan size categories defined earlier. Each vector was then correlated with the data, people, and things variables. The results of this exercise are shown in Table 3.9 where

Table 3.9. Correlations between vectors of occupational location quotients and occupational task ranks for CMSA/MSA population size categories

CMSA/MSA size category	Data	People	Things	Things (less category 7)
More than 5 million	−0.36**	−0.17**	0.20**	−0.23**
1 million–5 million	−0.27**	−0.15**	0.12*	−0.19**
500 thousand–1 million	0.10	−0.01	−0.10*	−0.33
250 thousand–500 thousand	0.27**	0.17**	−0.16**	0.15*
50 thousand–250 thousand	0.47**	0.27**	−0.20**	0.28**

*Significant at 0.05 level; **significant at 0.01 level.

the overall pattern of coefficients reconfirms our basic hypothesis with unusual clarity. The correlations related to the data variable demonstrate with strong statistical significance that occupations in large metropolitan areas are oriented to highly skilled forms of data manipulation (synthesizing, coordinating, etc.), whereas occupations in smaller metropolitan areas are much less skilled in this regard (copying, comparing, etc.). Likewise, large metropolitan areas score highly on occupations that call for sophisticated forms of personal interaction (mentoring, negotiating, etc.), but smaller metropolitan areas are more focused on relatively commonplace forms (serving, taking instructions, etc.). Once more, the role of the things variable only makes sense after we eliminate category 7, and when this is done, we see that large metropolitan areas are strongly associated with occupations that involve significant skills (setting up, precision working, etc.), whereas small metropolitan areas are much more likely to harbor routine production jobs (tending, feeding–offbearing, etc.). It is of some interest to note from Table 3.9 that metropolitan areas of intermediate size (i.e. between 500 thousand and 1 million people) evidently have precisely intermediate occupational profiles between those of large and small metropolitan areas.

A slightly different but entirely consistent perspective on the same data is given in the correlogram represented in Figure 3.1. This figure is based on simple correlations between the vectors of location quotients (as defined in the previous paragraph) for each of the five metropolitan size categories, designated here C_1, C_2, C_3, C_4, and C_5, in order of descending population size. Each of the main lines in Figure 3.1 joins up the coefficients of correlation between a given size category of metropolis and the other four. It is immediately evident that the two largest size categories (C_1 and C_2) behave in a fashion

59

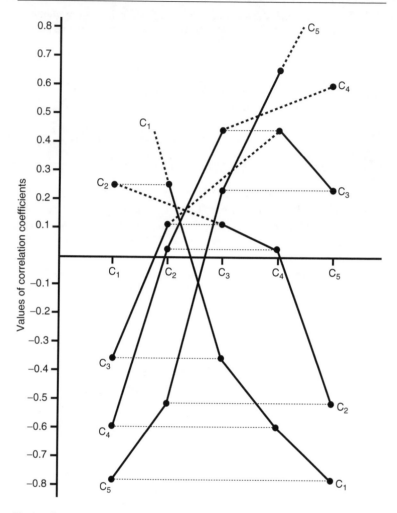

Figure 3.1. Correlogram showing relations between different CMSA/MSA population size categories in regard to occupational structure. Each point in the diagram represents the correlation between CMSA/MSA size categories i and j using location quotients for 362 different occupations as data. The CMSA/MSA population size categories are more than 5 million (C_1), 1 million to 5 million (C_2), 500 thousand to 1 million (C_3), 250 thousand to 500 thousand (C_4), and all other CMSAs/MSAs (C_5). Heavy dashed lines represent interpolations across loci where correlation coefficients are of the type r_{ii} and hence equal to unity. Dotted horizontal lines link points representing dyads of equivalent correlation coefficients, that is r_{ij} and r_{ji}

that is diametrically opposite to the smallest two (C_4 and C_5), while the middle category (C_3) again represents an intermediate trend. In other words, large metropolitan areas and small metropolitan areas appear to be two rather distinctive worlds in regard to the detailed occupations that gravitate to the one or other milieu. If we go back to the original data on which Figure 3.1 is based, we find again that occupations in large metropolitan areas are characterized by more complex tasks on the data, people, and things spectrum, while occupations in small metropolitan areas are much more likely to be concerned with more routine and standardized tasks.

The Large Metropolis and the Cognitive-Cultural Economy in the Global Era

Two main sets of empirical findings emerge from the above discussion. The first is that a very general pattern of functional relationships can be observed such that the economies of large metropolitan areas are rather clearly specialized in production activities marked by shifting, open-ended, unstable production relations and a labor force that exhibits high levels of valency with respect to the tasks that it is called upon to perform. By contrast, small metropolitan areas are more prone to be associated with relatively regimented production activities embodied in large and capital-intensive establishments. The second and related finding is that the emerging cognitive-cultural economy of the United States is being ushered in above all via metropolitan areas at the top of the urban hierarchy. In particular, large metropolitan areas are places in which significant concentrations of skills in regard to data-oriented, people-oriented, and things-oriented jobs can be found. To be sure, large metropolitan areas are also places where low-wage manufacturing and service jobs abound, but these types of jobs (e.g. in fashion industries, vehicle operation, restaurant trades, hotel service, and so on) are not only deeply and increasingly intertwined with the main cognitive-cultural economy but also in their turn dependent on definite kinds of cognitive and cultural talents. Even so, there are exceptions to the general finding that smaller metropolitan areas tend to be relatively averse to cognitive-cultural economic activities. As the examples of Austin in Texas (high-technology industry), Nashville in Tennessee (music), and Las Vegas or Reno in Nevada (tourism and entertainment) indicate, a significant number of special cases can be observed where the cognitive-cultural economy flourishes even in urban areas of relatively modest size. Small university towns, too, like Ann Arbor or State College, are places where the quotient of cognitive-cultural workers is abnormally high relative to other small centers.

Large metropolitan areas, of course, have always been places in which a relative abundance of creative, innovative, and resourceful individuals are to be found (Hall 1998). At the present time, the burgeoning of the cognitive-cultural economy is greatly intensifying this condition, and this circumstance is also shaping the human and physical assets of the large metropolis in quite distinctive ways. Large cities, not only in the United States but across the globe, are going through many peculiar transformations as these processes work themselves out. These changes are not confined to the economic sphere, but are also visible in the restratification of metropolitan society that is currently under way, and in the changing physical appearance of the city as gentrification in all its different guises gives rise to widening landscapes of urban redevelopment. Many of these changes in the urban environment intertwine with the cognitive-cultural economy to bring forth new and distinctive systems of competitive advantages in large metropolitan areas. In view of this proposition, and in contradistinction to the conclusions of Kim (1995) in regard to an earlier phase of American urban development, we may infer that localized external economies and increasing-returns effects (as opposed to natural endowments and internal economies) constitute a major and intensifying factor in the resurgence and economic differentiation of metropolitan areas in the United States today.

Appendix: Variables Used in the Regression Analyses and their Definitions in Terms of *Economic Census* Categories

Average salary: Annual payroll minus ($1,000) minus production workers' wages ($1,000) divided by total employment minus average number of production workers a year.

Average size of establishments: Total employees divided by total number of establishments.

Capital–labor ratio: Beginning of the year gross value of assets ($1,000) divided by average number of production workers a year.

Hourly wage: Production workers' wages ($1,000) divided by average number of production workers a year.

Inventory turns: Total value of shipments ($1,000) divided by total beginning-of-year inventories ($1,000).

Machinery rentals per production worker: Machinery rentals ($1,000) divided by average number of production workers a year.

Materials intensity: Total cost of materials ($1,000) divided by total value added ($1,000).

Percent production workers: Average number of production workers a year as a percentage of total number of employees.

Ratio of contracts to shipments: Cost of all contract work ($1,000) divided by total value of shipments ($1,000).

4

The Cognitive-Cultural Economy and the Creative City

Economy and Society

The economic order of capitalism has always been deeply inflected by purely social, noneconomic forces and this has perhaps never been more obviously or widely apparent than at the present moment in history when so much of the domain of production, work, and exchange intersects with the more subjective and personal dimensions of the cognitive and the cultural. Some of the most visible points of junction between these moments of contemporary life are to be discovered in the modern city with all of its multifarious forms of spatial and functional proximity. At the same time, the deepening cognitive-cultural character of the economy of many cities has been attended by a number of rather surprising outcomes in the social and physical aspects of intra-urban space.

Much of the sense in which I am using the expression "cognitive-cultural" here can be grasped initially by noting that significant elements of the sphere of productive activity today thrive on scientific knowledge inputs, continuous innovation, product multiplicity and differentiation, the provision of customized services, symbolic elaboration, and so on. The notion of a cognitive-cultural economy refers above all to the circumstance that as these developments have come about, labor processes in general have come more and more to depend on intellectual and affective human assets. Equally, in selected segments of the economy, labor processes are becoming less and less focused on minutely partitioned and routinized forms of work. To be sure, the cognitive-cultural dimension has always been more or less present in earlier versions of capitalism, even in the highly mechanized system of fordist mass production. In the fordist economy, however, producers generally faced the essential impera- tive of reducing costs by means of large-scale mechanization, and to assimilate—as far as possible—the labor force into a regimen of work that consisted primarily of a set of simple physical operations (cf. Braverman 1974). In today's cognitive-cultural economy, by contrast, the rational and emotive faculties of the labor force are being dramat- ically revalorized in the workplace, and these faculties are becoming increasingly requisite even in the case of jobs at the lower end of the wages spectrum.

Cognitive-Cultural Capitalism

Any concrete expression of capitalist economic order can typically be described in the first instance by reference to (*a*) its leading sectors, (*b*) its technological foundations, (*c*) its conventionalized structures of labor relations, and (*d*) its characteristic forms of competition and market order (Boyer 1986). These individual activity systems are man- ifest in unique ways in the cognitive-cultural version of capitalism that is emerging in many different quarters today. Let us consider each of them in turn.

- Much of the contemporary economy is being driven forward by key sectors like technology-intensive manufacturing, services of all varieties (business, financial, personal, etc.), fashion-oriented neo-artisanal production, and cultural-products industries. These sectors by no means account for the totality of the capitalist production system at the present time, but they are assuredly at the leading edges of growth and innovation in the most econom- ically advanced countries.

- Notwithstanding the evident heterogeneity of these sectors, they have all been deeply penetrated by digital technologies that have in turn facilitated the widespread deroutinization of labor processes and the destandardization of outputs.

- Employment relations have been subject to radical flexibilization and destabilization, thereby injecting high levels of precarious- ness into labor markets for workers at virtually all levels of skill and human capital formation. Also, project-oriented work based

65

on shifting teams of individuals, each of whom brings distinctive skills and talents to the labor process, has become increasingly important (Grabher 2002, 2004). Project-oriented work itself can be seen as extending across an organizational spectrum that passes through various forms of integration and quasi-integration of work teams in individual firms to full-blown interfirm networks as found in industries like motion pictures.

• A marked intensification of competition has occurred (reinforced by globalization) in all spheres of the economy, though much of this competition occurs in modified Chamberlinian form because products with high quotients of cognitive-cultural content often possess quasi-monopoly features that make them imperfect substitutes for one another, and hence susceptible to niche-marketing strategies.

As these trends have moved forward, the old white-collar/blue-collar principle of productive organization and labor-market stratification so characteristic of classical fordism has also been deeply modified. Autor et al. (2003) and Levy and Murnane (2004) have argued that the advent of computerization has meant that many of the routine functions that were integral to the work of both the old white-collar fraction (e.g. accounting, records management, calculating, information sorting, and so on) and the old blue-collar fraction (repetitive manual operations above all) are rapidly disappearing. Even many segments of the economy where digital automation is not economically feasible are declining in most of the more advanced capitalist societies as low-wage standardized jobs shift increasingly to cheap-labor locations. In this sense, digitization and delocalization can be seen as partial substitutes for one another.

Wherever the cognitive-cultural economy is developing apace—above all in major metropolitan centers of the world system—a new overarching division of labor appears to be overriding the older white-collar/blue-collar split. On one side of this new arrangement we find an expanding core or elite labor force whose work is concentrated primarily on high-level problem-solving and creative tasks. On the other side we find a new peripheral fraction of the labor force that is increasingly called upon to function in jobs like flexible machine operation (e.g. a driving a vehicle or manipulating a sewing machine), materials handling (e.g. small-batch assembly of variable components), security functions, janitorial tasks, personal services, and so on. Although these jobs may involve significant degrees of physical engagement and call for much less in the way of formal qualifications and training than jobs in the upper tier, they too are often imbued with varieties of meaningful cognitive-cultural content if only because workers must have a rather well-developed sense of how to proceed from one discrete operation to the next, and, in many instances, how to deploy certain affective-behavioral attributes, for example in many segments of the new service economy (McDowell 1999). In any case, the usual blanket designation of these kinds of jobs as "unskilled" is quite contestable.

The upper tier of the labor force of the cognitive-cultural production system can be further identified in terms of broad occupational categories like managers, professional workers, business and financial analysts, scientific researchers, technicians, skilled craftsworkers, designers, artists, and so on. These are occupations that require significant levels of human capital, and they are generally well paid, though certainly not invariably so (McRobbie 2004). They are, in any event, essential drivers of efficacy and competitive proficiency in the modern economy. First, managerial and allied workers carry out the functions of administration, monitoring, and control of the production system as a whole. Second, skilled analysts and other professionals are needed to maintain the specialized business and financial operations of modern capitalism. Third, scientific and technical workers are employed in large numbers to supervise the underlying technological infrastructure of the cognitive-cultural economy as well as to satisfy its unquenchable thirst for high levels of innovation. Fourth, many of the most dynamic sectors of the cognitive-cultural economy are characterized by a strong service element requiring skilled human intermediation at the producer–consumer interface. Fifth, workers with well-honed artistic and intellectual sensibilities make up an increasingly important part of the labor force, for contemporary capitalism is also the site of a remarkable efflorescence of cultural-products industries in the broadest sense, that is industries whose final outputs are permeated with at least some degree of aesthetic and semiotic content, and where such matters as fashion, meaning, entertainment value, look, and feel, are decisive factors in shaping consumers' choices about the products that they buy. In each of these types of employment, heavy doses of the human touch are required for the purposes of management, research, information gathering and synthesis, communication, interpersonal exchange, design, the infusion of sentiment, feeling, and symbolic content into final products, and so on. As noted earlier, the elite labor force that sustains these functions of the cognitive-cultural economy is expanding rapidly at the present time, especially in major metropolitan areas.

The lower tier of the labor force in the cognitive-cultural economy is employed in a thick stratum of manual production activities that are not only less well remunerated but also generally much less gratifying in their psychic rewards. I am referring here to both the workshop and factory operations that are part and parcel of the contemporary cognitive-cultural economy (in many high-technology and neo-artisanal sectors, for example), as well as to jobs in services such as janitorial and custodial work, facilities maintenance, low-grade hotel and restaurant trades, and so on. Additionally, a significant informal employment niche is sustained by the demands of more highly paid workers for domestic labor to perform tasks such as housecleaning, repair work, gardening, and childcare. This extended underbelly of the cognitive-cultural economy is notorious for its sweatshop operations and frequent brushes with illegality in regard to labor laws. In the more advanced countries, a high proportion of the labor force in this segment of the production system is made up of immigrants (many of them undocumented) from developing parts of the world. Large numbers of these immigrants form a polyglot underclass with at best a marginal social and political presence in their host environments.

The gap between the average incomes of the two strata of the workforce identified in the previous paragraphs has been growing apace in the United States over the last decade or so (Autor, Katz, and Kearney 2006; Morris and Western 1999; Yun 2006). In addition, both strata are subject to much labor-market instability. Workers of all types face increasingly frequent bouts of unemployment, and are more and more likely to be caught up in temporary, part-time, and freelance modes of labor. Along with these shifts in the structure of the employment relation has gone what some analysts identify as a declining sense of allegiance among workers to any single employer (Beck 2000). To be sure, the capacities of each of the two main strata of the cognitive-cultural labor force for dealing with these predicaments differ dramatically. While social networks are a major source of labor-market information for both groups, individuals in the upper stratum usually command significant resources in terms of contacts and interpersonal know-how that allow them a far greater range of maneuver. In contemporary society, it is not uncommon to come across cognitive-cultural workers who have carried networking to something like a fine art, or more accurately, perhaps, a semi-routinized habit of life in which they devote considerable amounts of time to socializing with fellow workers and exchanging information with one another about job opportunities and the state of the labor market. Reputation is a key item of currency in these fluid employment conditions, and is a major factor lubricating the progress of upper-stratum workers through the employment system. An essential strategy deployed by many individuals in this stratum involves the accumulation of personal portfolios of employment experiences demonstrating the depth and diversity of their career paths and creative accomplishments hitherto (Neff, Wissinger, and Zukin 2005). For these workers, too, elaborate self-management of careers replaces the bureaucratized personnel functions of the traditional corporation, a state of affairs that results in a further set of uncertainties and perplexities for many individuals.

The Cognitive-Cultural Economy and the Metropolitan Milieu

The Changing Urban Economy

As this new economic order of things has gathered steam over the last couple of decades, it has come to ground preeminently, but by no means exclusively, in large metropolitan areas, and most especially of all in major global city-regions like New York, Los Angeles, London, Paris, and Tokyo (Sassen 1994). These are the flagship hubs of the new economy, and the primary nerve centers of a cognitive-cultural production system increasingly geared to markets that extend across the entire globe.

Cognitive-cultural production activities, then, are typically highly concentrated in geographic space, yet their market reach often extends to the far corners of the world. Two analytical lines of attack help to clarify this apparently paradoxical state of affairs. In the first place, many types of producers in cognitive-cultural sectors of the economy have a definite proclivity to agglomerate together in geographic space by reason of the increasing-returns effects residing in the external economies of scale and scope that flow from selected aspects of their joint operation in particular localities. The role of flexible interfirm networks, local labor markets, and localized processes of learning and socialization is especially critical here (Cooke and Morgan 1998; Scott 2000a; Storper 1997). Groups of producers with strong interdependencies in regard to these variables have a powerful inducement to gravitate toward their common center of gravity, thereby reducing the space–time costs of their traded and untraded transactional interrelations and enhancing the total stock of jointly generated external economies. The inducement to agglomeration is heightened by the market instabilities that cognitive-cultural sectors commonly face. In the second place, even though it is true that low (and ever falling) transactions costs make it possible for certain kinds of firms to dispense altogether with the advantages of agglomeration and to decentralize to low-cost locations, the same phenomenon also permits many other kinds of producers to enjoy the best of both worlds, that is to remain anchored within a specific cluster, and thus to continue to appropriate localized competitive advantages while simultaneously contesting far-flung markets.

As the market range of producers in any given cluster increases, local economic growth accelerates, leading to the deepening of localized increasing-returns effects and the intensification of agglomeration. The signs of this developmental dynamic are palpable in the world's great metropolitan areas today, both by reason of the rapidly growing incidence of cognitive-cultural sectors in their overall economic structure, and in the frequent expression of this growth in the formation of intra-urban industrial districts devoted to specialized facets of cognitive-cultural production (see, for example Arai, Nakamura, and Sato 2004; Currid 2006; Rantisi 2004; Schoales 2006). Classical examples of such developments are high-technology and software production in the San Francisco Bay Area, the entertainment industry in and around Hollywood, the business and financial centers of New York and London, and the fashion worlds of Paris, Milan, and Tokyo. In many instances, any given metropolitan area will contain diverse clusters of sectors like these, as illustrated by Figure 4.1 which shows some of the more important industrial districts that constitute the basic spatial scaffolding of the cognitive-cultural economy of Los Angeles. Observe in this example that districts composed of more labor-intensive sectors such as clothing and film production

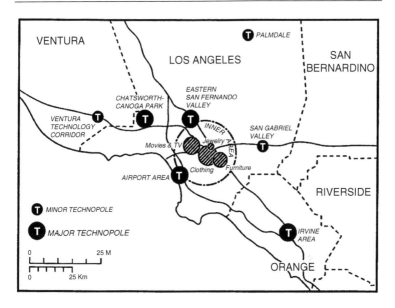

Figure 4.1. Selected industrial districts in Los Angeles and adjacent counties

occupy relatively central locations, whereas districts containing more technology-intensive industries are found in discrete technopoles scattered around the fringes of the metropolitan area. These contrasts are no doubt a function of the differing transactions costs (in terms of both interfirm relations and labor-market interactions) of these two kinds of economic activity relative to their requisite land inputs, but there is, as far as I can ascertain, no established analytic model that identifies the precise relationships that bring about this sort of situation.

The Changing Social and Physical Milieu of the City

Along with the widespread growth of cognitive-cultural production systems in the modern city have come numerous parallel transformations of the social and physical attributes of intra-urban space, including significant aesthetic enhancements of privileged parts of the urban fabric. Among the most symptomatic expressions of these trends is a general process of socioeconomic upgrading in downtown areas and surrounding inner city areas. This process

71

is widely referred to in the literature as "gentrification" (Smith 2002), though the term leaves much to be desired. In fact, it was originally coined in relation to a more limited set of phenomena, focused specifically on incursions of middle-class households into decaying inner city neighborhoods (Glass 1963). What is at stake in the gentrification process nowadays is nothing less than radical and continuing transformations of extensive portions of the city not only by upward social transformation but also by a twofold logic of cognitive-cultural economic development and the reimaging of significant parts of intra-urban production space and social space by means of dramatic new architectural symbologies.

An increasingly common manifestation of this process is the recycling and upgrading of extensive old industrial and commercial zones of the city to provide new spaces able to accommodate high-level production and consumption activities. Harbor Front in Baltimore, Docklands in London, and the Zürich West development in Switzerland are outstanding instances of this phenomenon. Similar kinds of initiatives can be found in Britain in Manchester's Northern Quarter and Sheffield's Cultural Industries Quarter with their aspirations to develop as dynamic hubs for small creative enterprises such as recording studios, electronic media labs, fashion design activities, and so on. In Los Angeles, to cite another example, a new Fashion District just to the south of the central business district has recently been created in what was originally a dispiriting cluster of grimy clothing factories. This development, with its renovated buildings and colorful street scenes, expresses the rising status of the Los Angeles clothing industry as a global center of designer fashions, and helps to sustain the newfound ambitions of many local producers to compete in high-end markets. An analogous type of development is observable in the so-called SOMA[1] area of San Francisco where a neighborhood of decaying commercial and residential buildings has been transformed over the last decade or so by incursions of new media producers. In parallel with initiatives like these, local authorities in cities all over the world are more and more engaged in projects that involve the conversion of derelict facilities to serve a diversity of economic and cultural purposes, as in the case of Amsterdam's Westergasfabriek

---

[1] i.e. south of Market Street.

or parts of the Ruhr region of Germany where efforts to rebuild a decaying industrial landscape are aggressively under way.

A related and increasingly spectacular case of the recycling of urban space can be observed in the construction of large-scale architectural set pieces, functioning as iconic expressions of local economic and cultural aspirations in an age of cognitive-cultural capitalism. The grand projects set on foot by President François Mitterand in Paris in the 1980s represent one of the pioneering and certainly one of the most determined instances of this kind of ambition, and have done much to add to the already celebrated reputation of Paris as the city of spectacle and as a global cultural reference point. Other illustrative cases of urban reimaging projects in pursuit of economic and cultural status are the Guggenheim Museum in Bilbao, Toronto's Harbourfront, and the Petronas Towers in Kuala Lumpur. These and analogous architectural gestures register a presence on the global stage while generating prestige and cachet that spill over into the wider urban communities in which they are located. Urban elites in all parts of the world are increasingly committed to the pursuit of projects like these in attempts to assert the ambitions and visibility of their cities as foci of cultural interest and economic promise in the new global order as well as to augment the attractiveness of these cities to inward investors and highly qualified migrants.

Alongside these changes, large swaths of low-income housing in central city areas have been subject to appropriation and recolonization by the affluent. This process is discernible both in the renovation of old working-class residential properties and derelict slum buildings and in wholesale land clearances to accommodate new blocks of expensive condominiums. Gentrification in this sense has actually been occurring in American cities for the past several decades, but it has accelerated greatly in recent years as a result of changing structural conditions in the urban environment and changing priorities in residential preferences. As jobs in traditional manufacturing and wholesaling activities have declined in inner urban areas, much of the old working-class population in adjacent communities has migrated to other parts of the city. Correspondingly, job opportunities for cognitive-cultural workers in and around the central business districts of large cities have mushroomed of late years, and many of these workers are taking up residence in nearby neighborhoods in order to reduce commuting times and to gain access to burgeoning shopping, leisure, and cultural facilities in the central city. Very often, the first sign that a dilapidated section of the inner city is destined to go through this sort of transition is the irruption of groups of artists and bohemians in the area and the blossoming of studios, cafés, clubs, and so on, serving their needs (Zukin 1982). Indeed, some analysts have accorded these groups, along with gays, a special status as key harbingers and tracking molecules of the "creative city" syndrome (cf. Florida 2004; Lloyd 2002; Lloyd and Clark 2001). The overt presence in the urban landscape of groups like these is also said to symptomatize a state of openness and tolerance in local society, qualities that are thought, in turn, to be essential for the blooming of a creative urban environment. As such, the presence or absence of these groups in the city is taken by some commentators to represent a sort of litmus test of local prospects for general "creativity," a concept that I shall subject to more intense scrutiny at a later stage.

There are numerous signs, then, of important shifts in the functions and form of the city as the cognitive-cultural foundations of modern capitalism have deepened and widened. These shifts are detectable in the economic patterns, social organization, and physical structure of many different cities. Specialized areas of the city dedicated to entertainment, recreation, edification, shopping, and so on, have also undergone much elaboration and embellishment as individuals with high levels of cognitive and cultural capital— not to mention pecuniary capital—have become a more insistent component of contemporary urban life (Zukin 1995). In these ways, a new kind of balance and integration seems to be emerging at least in privileged sections of modern cities between economy and society, between production and consumption, between work and leisure, and between commerce and culture. A dark shadow is nonetheless cast over this gratifying picture both by the swelling underbelly of low-wage industrial and service functions that are invariably to be found in large metropolitan areas where cognitive-cultural economic functions are most highly developed, and by the often problem-ridden residential areas, no matter where they may be located in the city, that are the sources of the labor needed to maintain these functions. The deepening pall cast by this condition of social and economic inequality almost certainly puts shackles on the capacity of the city to promote consistently high levels of social learning, economic innovation, and human conviviality. Large segments of the urban population face serious impediments to participation as full-blown citizens in daily life and work, a circumstance that generates high costs to the individuals directly concerned, and—via the multiple negative externalities that result from this situation—to urban society as a whole. The relentless withdrawal of public services that is occurring in the context of the rightward political shifts that are evident in many of the more advanced capitalist countries at the present time only serves to intensify the possessive individualism characteristic of so much of modern urban life at the expense of more communal values.

Cognitive-Cultural Workers and the Constitution of Urban Life

Various intimations of the meaning of the new economic forces and social alignments that are rising to the fore in capitalist society are now a common feature in both the journalistic and academic literature. Among the more prominent of these effusions on the new economy is a stream of managerial theories and advice concerned with the personal and affective qualities required to bring order and dynamism into the cognitive-cultural workplace. The normative discourse of management analysts and consultants today is considerably less concerned than it once was with down-to-earth issues of efficiency and control, and much more focused on methods of cultivating human resources like leadership, empathy, self-motivation, adaptability, inventiveness, resourcefulness, ethical consciousness, and so on, in a fast-moving, high-risk business environment (Boltanski and Chiapello 1999; Thrift 2005). There is incontestably much in this discourse that is helpful to managers and workers trying to find some sort of strategic purchase on the day-to-day problems that they face in the new cognitive-cultural economic environment. From all that has gone before, however, it is considerably less useful as a guide to the formulation of critical insights or as a basis for the construction of sensible and politically plausible imaginaries about alternative possibilities.

In parallel with these efforts to comprehend the nature of elite labor tasks in contemporary capitalism, many social scientists over the last few decades have sought to describe the changing stratification of modern society, and above all to typify the transformations that have been occurring in social structure since the heyday of the classical white-collar/blue-collar division that prevailed in American cities over much of the twentieth century. In a pioneering statement, Bell (1973) alluded to the advent of what he called postindustrial society, and he suggested that the old social divisions of capitalism were in fact being transcended by a newfound drive for personal fulfillment and self-realization in a service-oriented economy. Gouldner (1979) offers us the idea of a "new class," made up of individuals who have internalized an ideology of critical rationality, so that, for them, reasoned arguments take precedence over hierarchical authority as a basis for belief and action; the modern technocrat is the emblematic figure of this new class. Reich (1992), in turn, refers to "symbolic workers" who constitute, he claims, the elite of an emerging information society. French and Italian researchers have recently put forward the idea of a "cognitariat" that functions as a new type of privileged labor force in capitalism (Moulier Boutang 2007; Rullani 2000). Sklair (2000) broadens the picture with the concept of a "transnational capitalist class" composed of managers, professionals, technicians, and so on, engaged in forms of work that express and promote the historical project of globalization. And Florida (2002) has advanced the argument that a new "creative class," comprising all those workers engaged in one form or another of thought-intensive work, has come into being in American society.

Each of these attempts to capture elements of the changing organization of society in contemporary capitalism again has something of interest and significance to convey, though none can be deemed entirely satisfactory. The term "class," is perhaps unduly forceful a word to use for some of these rather nebulous social groupings, especially in view of its more orthodox connotation of two opposing strata whose interests clash as a consequence of their structured relations to the means of production and their opposing claims on the economic surplus. Conceivably, we might capture something of current social divisions in the more diluted Weberian idea of class with its emphasis on occupation and relative life chances. Even so, as Markusen (2006) has argued, Florida's proposed creative class is something of an incoherent concept, for it assembles a wide assortment of very disparately situated individuals—from company executives to struggling artists and from international financiers to school teachers—within its rather elastic boundaries. What can be said with assurance in this context is that a marked restructuring of urban society has been proceeding over the last few decades, leading in one sense to unusually high levels of urban social variegation, but also and in another sense to an ever intensifying polarization between the top and the bottom ends of the labor market (see Chapter 6).

Much current writing on so-called creative cities is especially problematical because it invests those individuals who compose the more privileged segments of contemporary capitalist society with a sort of ontological capacity for "creativity," a characterization that carries with it an overload of exhilarating implications, but that is also rather threadbare in terms of its concrete meaning. In reality, the distinctive forms of human capital that these individuals possess and the specifically *cognitive and cultural* tasks they are called upon to perform in the routines of daily work are for the most part wedged in definite social grooves and infused with very specific substance, not all of which, incidentally, can be taken as an unmixed blessing. Within the framework of contemporary capitalism, these tasks are focused on activities like neoliberal technomanagement, innovation-oriented process retooling and product design, the personalized provision of services, the naturalization of socially useful aptitudes and beliefs (in educational institutions and the media, for example), and the commercialization of experiences, cultural encounters, leisure pursuits, and so on. Special mention needs to be made here of the enormous recent expansion of cultural-products industries generally and the concomitant emergence of an important segment of the cognitive-cultural labor force dedicated to the conception and fabrication of outputs whose function is to entertain, to instruct, to embellish, to reinforce identity, and so on (Bourdieu 1979; Hesmondhalgh 2002; Power and Scott 2004). Equally to the point, creativity is far from being an extra-social abstraction (that has somehow, miraculously, proliferated in recent years), but rather is a socially constructed phenomenon. Just as creativity in nineteenth-century Lancashire revolved to a significant extent around the search for improvements in cotton textile production, and creativity in Detroit in the mid-twentieth century was concerned above all with innovation in the automotive sector, so creativity in today's cognitive-cultural economy is not an ethereal endowment that comes somehow from outside (and in some versions, literally, from outside the city and on the backs of high-level migrants) but a materially grounded reflection of the challenges and opportunities that are increasingly opening up to workers at the present time, and nowhere more so than in large metropolitan areas of the developed world.

By the same token, the lives and consciousness of these workers take shape in very concrete circumstances, among which the rising levels of general social instability and risk represent a problem of special importance (Beck 1992, 2000). Even the urban elite is subject to an intensification of the general precariousness of life. Individual members of the labor force accordingly exert considerable energy and time in navigating pathways through the reefs and shoals of practical social existence, whether by means of self-conscious social networking on the part of upper-tier workers (Batt et al. 2001; Neff, Wissinger, and Zukin 2005; Ursell 2000), or via diverse ethnic and extended-family ties on the part of the lower tier (Sanders, Nee, and Sernau 2002; Waldinger and Bozorgmehr 1996). Many kinds of cognitive-cultural workers—especially in the early stages of their careers—are inveterate joiners of work-related social groups, and they are prone to spend large amounts of time outside their normal working hours in building relationships with allied workers so as to maintain their labor-market edge. In these conditions, human interaction is apt to take on discernible utilitarian undertones. Thus, in her study of workers in the television industry Ursell (2000) has shown how an "economy of favors" has arisen in which information about job opportunities and work-related matters is exchanged on an informal *quid pro quo* basis through extended webs of social contacts. At the same time, the kaleidoscope of shifting opportunities and setbacks that characterize labor markets in much of the cognitive-cultural economy today is increasingly reflected in careers that unfold across many different employers in many different places, and even— especially for upper-tier workers—in many different countries. In this manner, the traditional connection between propinquity and community is subject to further decay, just as a growing ethos of possessive individualism or interpersonal engagement without durable commitment becomes a normalized condition of urban existence. The same instability and insecurity provide a strong incentive for members of the upper tier of the labor force to engage in persistent self-promotion and self-publicity, an incentive that no doubt is magnified the more they are possessed of an individualized portfolio of experiences and qualifications that mark them out as the bearers of unique packages of attributes and talents. In testimony to the above remarks, Sennett (1998) has pointed to an apparent corrosion of traditional forms of affectivity and trust in both the workplace and social life, while Putnam (2000) has written more generally about the weakening of communal ties in America.

It is tempting to attribute at least some of the narcissism that was presciently thought by Lasch (1978) to be on the rise in the American psyche to social forces and predicaments of these types. A less ambitious way of making much the same point is to appeal to the accumulating evidence of the expansion of the sphere of the private and the personal, and a corresponding contraction of the public sphere in American cities. Quite apart from the condition of public penury and a broadly decaying sense of community, as already invoked, we can see the immediate effects of this state of affairs in the intense fragmentation of the social space of the contemporary metropolis. The very social diversity that is so often celebrated as one of the main conditions of a creative urban environment today is actually inscribed on the landscape of the metropolis in patterns of separation and detachment, accentuated by the striking marginalization of the ever-expanding immigrant population of the city. For many immigrants, this situation is manifest not only in their relative and absolute poverty but also in the political disenfranchisement to which they are subject. The fact that so many of these denizens of American cities in the early twenty-first century have curtailed entitlements and restricted channels for the democratic expression of their political aspirations means not only that they are denied full incorporation into urban society but also that they have limited incentives to make durable commitments to the community at large. The net result is surely a significant deterioration of the capacity of the urban system for releasing and mobilizing the creative potential of the citizenry at large.

Beyond the Creative City

As cognitive-cultural forms of production and work penetrate more deeply into contemporary capitalist society, enormously varied bundles of urban responses have been set in motion. On the one side, a set of privileged intra-metropolitan spaces supporting the work, residence, and leisure activities of the new cognitive-cultural elite is now an important ingredient of many world cities. On the other side, and given that large numbers of low-wage, low-skill jobs are a major element of the cognitive-cultural economy, a growing underclass is also an insistent feature of the very same cities. These trends are embedded in a widening dynamic of economic-*cum*-cultural integration at the world scale, leading to complex forms of urban specialization and interdependence across the global landscape.

Some of the more positive features of this picture have of late been highlighted in a number of normative commentaries focused on the creative potentials of contemporary cities. Policymakers and planners in many different parts of the world have understandably displayed much enthusiasm in regard to these commentaries, and in numerous cases have actually embarked on attempts to make their cities appealing to the talented and high-skill individuals who are thought, in the more prominent versions of the story, to be the *primum mobile* of the creative city. The idea of the creative city is all the more irresistible to policymakers in view of its promise of high-wage jobs in sectors of economic activity that are by and large environmentally friendly, and its alluring intimations about significant upgrading of the urban fabric. In a number of cases, practical attempts to pursue the idea have been complemented by efforts to mount displays of architectural master strokes designed to establish dramatized points of reference in the global race for economic and cultural influence.

Florida (2002, 2004) has been the most forthright instigator of a normative agenda of this general type, but his ideas find both implicit and explicit support in other work, including the "consumer city" concept as formulated by Glaeser et al. (2001) and the view of the city as an "entertainment machine" that Lloyd and Clark (2001) have proposed. Florida's suggested strategy for building the creative city can be schematized—with only a touch of willful skepticism—in terms of three main brush strokes. First, municipal authorities are advised to encourage the development of amenities that are claimed to be valued by the creative class. Bikeways and fashionable restaurants figure prominently in the suggestions offered here, and regression analysis seems to suggest that warm winters also help things along, though cinemas and art galleries are apparently of much less consequence. Second, Florida proposes that once appropriate packages of amenities are in place in any given city, members of the creative class will then be inclined to take up local residence, especially if, in addition, an atmosphere of tolerance and openness prevails in the area. Third, the dynamism of the local economy will presumably accelerate as a result of this inflow, while upscaling of the built environment and general enhancement of the prestige-*cum*-attractiveness of the city as a whole can also be expected to occur.

The key analytical maneuver in all of this revolves around the implication that cities are really only aggregates of free-floating individuals and that they can therefore be restructured simply by appropriate fine-tuning of their amenities so as to keep in place and to attract targeted population groups. Unfortunately, there is much in this kind of analysis that recalls the obdurate tautologies of neoclassical economics. For example, if we observe a significant tendency for individuals of type x to live in (or migrate to) places with attributes of type y the suggestion is then that the same individuals must have a "revealed preference" for y. Revealed preference accordingly explains their residence in (or migration to) places with these attributes. How do we know this? Because they live in (or have migrated to) those places! In a world that is driven only by individual decision-making and behavior and that is devoid of any structural logic this sort of account might be taken at its face value. But what if the suggested attributes (like sunshine) have a merely incidental relationship to this logic? It is important to point out here, right at the start, that cities are almost always subject to path-dependent growth trajectories in which both the supply and the demand for labor move in patterns of mutually cumulative causation. The primary engine of this process is not the unilateral accumulation of a particular type of labor force in any given place, but the apparatus of the urban production complex, that is the network of interrelated industrial and service activities generating locationally polarized economic development in the first instance. This type of developmental engine was rather obviously at work in earlier periods of capitalism, and on due consideration it is still detectable as the major motive force of urbanization in cognitive-cultural capitalism today.

Consider the case of factory towns in nineteenth-century England. It was not the prior massing together of dense working-class populations that explains the formation of these towns, even though the presence of a working-class population is essential for a factory town to function. In the same way, the growth of Silicon Valley in the second half of the twentieth century is not to be accounted for by invoking the prior existence of some undifferentiated creative class nearby, just as it would surely be absurd to claim that the driving force of the Valley's long-term expansion can be ascribed to continual incursions by members of that class in order to satisfy their revealed (?) preferences for amenity value. On the contrary, the historic accumulation in Silicon Valley of a labor force comprising a very specific fraction of the labor force made up of semiconductor technicians, computer scientists, software engineers, and so on is comprehensible only when we set this trend in the context of an evolving web of specialized production activities and employment opportunities tied in to ever widening final markets for semiconductors, computers, and software. Yes, the supply of labor is a crucial moment in the chain of temporal intermediations through which cognitive-cultural centers of production and work evolve, but it remains a subordinate moment in the sense that the generative power of local economic development resides preeminently in the path-dependent logic of production, agglomeration, and regional specialization. By the same token, dissipation of that power is a virtually inevitable road to ruin, even where large numbers of workers with high levels of human capital reside in the local area. Policymakers neglect these aspects of the problem at their peril.

In addition to the analytical flaws that underlie much recent work on the creative city, an odd reticence can be detected in many of the associated policy claims that have been advanced about the possibilities for revival of the social life and physical environment of cities by tapping into the expansionary powers of the cognitive-cultural economy. While it is certainly correct to suggest that cognitive-cultural forms of production and work offer new and dynamic possibilities for urban regeneration, it bears repeating that there is a dark side to the developmental dialectic of contemporary cities, and that the currently deepening trend to neoliberalism in basic economic and political arrangements is actually exacerbating the problem. This comment raises political issues about the reconstruction of urban society that go well beyond simple pleas for openness, tolerance, and diversity, which are no doubt excellent qualities in their own right, but that do not in any sense guarantee transcendence of social isolation, fragmentation, and inequality. To the contrary, even if these qualities were universally present, the ingrained structural logic of the contemporary economic and social order would still in all probability give rise to conspicuous inequities and injustices in large cities.

In contrast with the neoliberal political agenda that currently holds sway in the United States, and that is endemically associated with high levels of urban poverty and deprivation, only some sort of conscientious program of social democracy with a strong focus on redistribution, decent jobs for all, and the reengagement of the citizenry in relevant forums of political interaction seems to be at once feasible and sufficiently well armed to deal with the tasks of social reform implied by these remarks. Above and beyond the implementation of elementary principles of equity, justice, and participatory democracy, an additional challenge looms ahead. As cities shift more and more into cognitive-cultural modes of economic activity, the search for meaningful forms of solidarity, sociability, and mutual aid in everyday work and life becomes increasingly urgent, not just because these attributes are important in their own right but also because they help to enlarge the sphere of creativity, learning, innovation, social experimentation, and cultural expression, and are hence essential for the further economic and cultural flowering of contemporary cities.

Finally, an even broader social imperative is brought to the fore as the cognitive-cultural economy continues its ascent and as the symbolic-affective content of final outputs becomes ever more pervasive. Consumption of these outputs has potent direct and indirect impacts on human consciousness and ideological orientation, and this process, by the same token, generates massive externalities for all. These externalities give rise to complex dilemmas in society for they reappear in various social and political guises with deep implications for modes of social being. And precisely because they are externalities, their costs and benefits can never be adequately processed via market rationality alone. A vigorous cultural politics in at least the minimal sense of persistent public debate and mutual education about the personal meanings and political consequences of the consumption side of the cognitive-cultural economy—and about the possibilities of more critically informed participation in its development—is thus a further prerequisite of a progressive and democratic social order in contemporary capitalism.

5

Culture, Economy, and the City

A special claim on our attention is now made by the particular segment of the cognitive-cultural economy that is constituted by the cultural or creative industries. These industries represent an especially rich and problematical domain of inquiry, not only because their outputs embody significant aesthetic and semiotic content but also because they are expanding at a rapid pace in large metropolitan areas at the present time and are a source of much new urban growth. At the same time, they function as channels of the commodification of culture, a circumstance that, in its turn, raises many puzzling ideological and political concerns.

Modern cultural industries can be broadly represented by sectors that produce outputs whose subjective meaning, or, more narrowly, sign-value to the consumer, is high in comparison with their utilitarian purpose. Cultural industries can thus be identified in concrete terms as an ensemble of sectors offering both manufactured products and services through which consumers construct distinctive forms of individuality, self-affirmation, and social display, and from which they derive entertainment, edification, and information. Fashion clothing, jewelry, furniture, television-program production, recorded music, print media, and so on are all instances of the types of industries that are involved in the modern cultural economy. It is accordingly evident that the cultural economy constitutes a rather incoherent collection of sectors, though for our purposes, they are bound together as an object of inquiry by three important common features. First, they are all concerned in one way or another with the creation of sign-value (Baudrillard 1968) or symbolic value (Bourdieu 1971). Second, they are generally subject to the effects of Engels' Law, meaning that as disposable income expands, consumption of these outputs rises at a disproportionately higher rate. Third, they exemplify with special force the dynamics of Chamberlinian or monopolistic competition. We might add a thought from the previous chapter to the effect that they are also frequently subject to economic pressures that encourage individual firms to agglomerate together in dense specialized clusters or industrial districts, yet their products circulate with increasing ease on global markets.

It must be stressed at once that there can be no hard and fast line separating industries that specialize in purely cultural products from those whose outputs are purely utilitarian. On the contrary, there is a more or less unbroken continuum of sectors ranging from, say, motion pictures or recorded music at the one extreme, through an intermediate series of sectors whose outputs are varying composites of the cultural and the utilitarian (such as shoes, kitchen utensils, or office buildings), to, say, iron ore and wheat at the other extreme. One of the peculiarities of modern capitalism is that the cultural economy continues to expand at a rapid pace not only as a function of the growth of discretionary income but also as an expression of increasing design intensiveness in ever-widening spheres of productive activity as firms seek to improve the styling of their outputs in the unending search for competitive advantage (Lawrence and Phillips 2002).

Cultural industries have therefore been significantly on the rise of late, and they are notably visible as drivers of local economic development at selected locations. Even such unlikely places as certain old manufacturing towns in the Midlands and North of England (Wynne 1992) or the German Ruhr (Gnad 2000), once widely thought of as representing quite inimical milieux for this type of enterprise, are now selectively blooming as sites of cultural production. Many authors, indeed, have commented of late on the potentialities of the cultural economy for job creation and urban regeneration in stagnating areas (see, e.g., Bassett 1993; Bianchini 1993; Bryan et al. 2000; Fuchs 2002; Hudson 1995; Landry 2000; Lorente 2002; O'Connor 1998; Throsby 2001; Whitt 1987), and as we saw in the preceding chapter, this kind of advocacy has taken on special force as the idea of the creative city has been increasingly adopted in policy circles. The present chapter is a further attempt to offer some critical guidelines about these matters via an assessment of the potentialities of cultural industries as instruments of urban and regional growth while simultaneously maintaining a judicious eye on the limitations and pitfalls that are likely to be attendant on any major policy thrusts in this direction.

A Digression on Aesthetics, Accumulation, and Urbanization

From its very historical beginnings in the seventeenth and eighteenth centuries, capitalism and the commercial values that go with it have been widely perceived as being fundamentally antithetical to many kinds of cultural interests, especially those revolving around aesthetic and artistic objectives. Culture, it is often thought, starts where the market ends. This incompatibility never seemed more complete than in the nineteenth century, when the economic order was represented by regimented and often dehumanizing forms of industrialization and urbanization, while much of the art of the period was enmeshed in otherworldly romanticism. Ruskin's *The Stones of Venice*, published right in the middle of the century, functioned as a paean to the past of Gothic crafts and architecture, and as an indictment of the cultural degradation that seemed to be deepening on all sides in industrial-urban Britain. Indeed, the very different social imperatives to which industrialization on the one hand and aesthetic practice on the other were subject at this time, not only set them in opposition to one another as a matter of principle but also seemed to put barriers in the way of their coexistence in close geographic proximity to one another. By and large, the lives of the proletarian workers and the factory owners who employed them were bound up with one set of urban conditions, while the lives of artists and their most enthusiastic audiences were bound up with another, and these conditions appeared more often than not to contemporary observers as being mutually exclusive. Of course, the incompatibility between the two sides was never absolute, and in some cities, various sorts of coexistence, if not interdependence, were worked out, as in the case of Chicago at the turn of the twentieth century where an intense industrial and commercial bustle prevailed in combination with a remarkable proclivity to architectural innovation and literary exploration. In certain metropolitan centers, most notably the Paris of Balzac and Zola, the worlds of industry, commerce, and culture came directly into contact with one another at selected points of social

86

and spatial intersection. Even so, places like the surging industrial cities of, say, northern England or the German Ruhr on the one hand, and the more traditional historical and cultural centers of Europe on the other, seemed to represent irremediably antithetical universes to many contemporary observers.

As the twentieth century progressed, this tension between accumulation and aesthetics continued to leave its mark on much of urban life and form in the advanced capitalist societies. With the shift of mass production to center stage of economic development, urban centers continued to grow and to expand outward in the Mammon quest. Many of the metropolitan areas of the American Manufacturing Belt in particular (Detroit, Cleveland, Pittsburgh, etc.) now appeared to numerous commentators as the archetypes of the utilitarian, philistine city. The mass-production system, indeed, carried the manufacturing and consumption of standardized commodities to new heights of intensity. To begin with, much of the output of the system consisted of consumer goods such as cars, domestic appliances, and processed foods, designed largely to absorb the wages of the burgeoning blue-collar workers who made up the majority of the urban populace. In addition, the system flourished on the basis of competitive cost-cutting, and hence (given its machinery-intensive structure) it was also marked by the routinization of manufacturing methods and the search for internal economies of scale in production. The system was thus endemically committed to the production of undifferentiated and desemioticized outputs, leading to the charges of "eternal sameness" that Frankfurt School critics were soon to level against its effects as it started to make incursions into popular culture (Adorno 1991; Horkheimer and Adorno 1972). The large industrial cities themselves were seen in many quarters as being given over to a syndrome of "placelessness" that critics like Relph (1976) ascribed to the increasing domination of technical rationality in mass society.

To be sure, Walter Benjamin writing in the mid-1930s had already set forth a series of quite hopeful views about the potentially progressive nature of what he called "mechanical reproduction" in the arts, and especially in cinema (Benjamin 1969). By the 1940s, however, the core Frankfurt School theorists were expressing grave concern about the application of industrial methods to the production of cultural outputs such as film, recorded music, and popular magazines. To people like Adorno and Horkheimer, these methods, being driven by capitalistic interests, were aesthetically suspect from the beginning, and they averred that the cultural content of the products to which these methods gave birth were patently manipulative and depoliticizing in practice.

It is nowadays fashionable to criticize the Frankfurt School theorists for their alleged *mitteleuropäischen* elitism, though the charge is blunted, perhaps, when their work is recontextualized within the conjuncture from which they wrote. Where they did err was not so much in the imputation of qualities of meretriciousness and triviality to commodified culture (such qualities are all too evident, then as now) but in their failure to see the possibility that the products of capitalist enterprise might also potentially be carriers of other more positive qualities. In one way or another, artistic culture is always produced in the context of definite historical and social conditions that in themselves lie outside the sphere of art but that shape aesthetic aspirations and practices. Provisionally, then, there is no reason to assert *on principle* that capitalist firms working for a profit are congenitally incapable of turning out goods and services with inherent aesthetic and semiotic value (or what Clive Bell (1924) in a rather different context, called "significant form").

Even as mass production was moving into high gear, a modernist aesthetic was already trying to come seriously to terms with its driving logic. Thus, as represented perhaps most dramatically by the Bauhaus and the great modernist urban design proposals of architects like Le Corbusier or Oscar Niemayer various aesthetic programs were mounted in theory (e.g. "less is more") and implemented in actuality in an attempt to give artistic expression to the main thrust of mass-production society (Banham 1960). More importantly for present purposes, the economic order of capitalism has evolved considerably since the days when members of the Frankfurt School were writing their pessimistic diagnoses of the cultural crisis of capitalism. The new cultural economy is capable of producing very much more variegated and inflected representational forms than the old mass-production system based on assembly-line methods. Concomitantly a sharp intensification of the economic significance and symbolic content of commodified culture has rather clearly been occurring of late. This trend is further reflected in an ever-increasing functional amalgamation of the spheres of culture and the economy, and—for better or worse—a discernible diminution of many of the strains and prejudices that formerly kept them in tense conflictual relationship to one another. I shall revisit some of these themes in the later discussion.

The Culture–Economy Nexus in Local Context

Selling Places

Certain places have always figured as destinations for tourists, pilgrims, scholars, and other visitors seeking entertainment, physical and spiritual recuperation, or cultural improvement. Over the eighteenth and nineteenth centuries various spas and resorts performed many of these functions for the upper echelons of European society, just as artistic centers like Florence and Venice exerted an enormous attractive force on visitors from far and wide. In the twentieth century a proliferation of places of this sort has come about, and especially in the later part of the century as diverse municipalities and other local authorities discovered the benefits of place-marketing and allied strategies to promote the commercialization of local historical, artistic, and cultural assets. We might refer to this phenomenon as a sort of first-generation policy approach to yoking culture and economy together in order to promote urban and regional prosperity.

Some time in the early 1980s, then, a rising awareness among policymakers and planners to the effect that the actual and latent cultural endowments of given places might be exploited in the interests of local economic development began to spread widely. This perception assumed high levels of plausibility in the policy arena in the light of multiplying practical achievements in the domain of place-marketing and the exploitation of local heritage assets for economic gains. Hitherto, urban and regional economic development programs had been greatly influenced by economic-base and growth-pole theories, which were seen as offering the most potent guidelines for salvation (cf. Perroux 1961). From this perspective, moreover, cultural assets and projects were taken to be almost entirely irrelevant. The one possible exception to the latter remark is the tourist industry, which, from an early stage, was sporadically extolled for its developmental possibilities in areas otherwise devoid of exploitable economic resources, (see, e.g., Wolfson 1967). The emergence of place-marketing and associated heritage-industry programs came greatly

89

into favor with entrepreneurial municipal governments over the 1980s (Harvey 1989) as many localities established different sorts of agit-prop agencies directed to improving their public image. These kinds of programs have continued to grow apace down to the present. In most cases, they have focused particularly on upgrading and redeveloping local cultural resources of all varieties (Graham, Ashworth, and Tunbridge 2000; Philo and Kearns 1993). Arts funding schemes are often deployed in combination with programs like these (Kong 2000; Williams 1997). One main objective, of course, is to attract increased numbers of visitors from other areas. Another is to enhance the image and prestige of particular places so as to draw in upscale investors and the skilled high-wage workers that follow in their train. These types of programs are also much in vogue as ways of encouraging urban regeneration, a feature that is exemplified in especially dramatic terms by numerous old industrial cities and regions that have sought to recycle deteriorated commercial and manufacturing properties as tourist, entertainment, and cultural facilities (Bianchini 1993).

Other types of schemes for advancing local visibility and generating income revolve—and increasingly so—around the promotion of festivals, carnivals, sports events, and similar mass spectacles (Gratton, Dobson, and Shibli 2001; Ingerson 2001). Local traditions and cultural idiosyncrasies offer a mine of exploitable possibilities here, as exemplified by the Bayreuth Wagner Festival, or the International Festival of Geography at St Dié-des-Vosges,[1] or New Orleans' Mardi Gras (Gotham 2002). In the same manner, the small Welsh market-town of Hay-on-Wye has parlayed its profusion of second-hand bookstores and its annual literary festival into a worldwide tourist attraction, thus confirming the point that while the most successful cognitive-cultural sectors may be concentrated in large cities, there are numerous specialized niches that can be served by producers in much smaller places. The success of Hay-on-Wye has encouraged numerous imitators in various parts of the world to follow its example (Seaton 1996). Another illustration of the conversion of local cultural peculiarities into visitor attractions is provided by Kinmen, Taiwan, where a long-standing arts and crafts tradition has been turned into a magnet for tourists (Yang and Hsing 2001).

---

[1] St Dié-des-Vosges is the town in eastern France where the early sixteenth-century cartographer Vautrin Lud carried out his work.

Nel and Birns (2002) describe an analogous case in Still Bay, South Africa, where the municipality has used place-marketing of its coastal location and climatic advantages to attract visitors, thereby overcoming a long history of economic stagnation. Perhaps the most remarkable instance of the remaking and marketing of place in recent years is furnished by the Guggenheim Museum in Bilbao, an initiative that has turned an old and stagnant industrial area into a world-renowned tourist center (Lorente 2002). Examples like these might be multiplied over and over again. The important point, however, is that while such first-generation policy strategies have achieved some notable successes, they are nonetheless greatly constrained as to both their range of applications and their likely economic results. Place-marketing strategies and allied methods of local economic development continue to be useful elements of the policymaker's toolkit, but they need to be put in due perspective, especially by comparison with a complementary set of approaches that has more recently started to come into focus. This remark points directly to an analytical question and a second-generation policy vision directed less to the selling of places in the narrow sense than to the establishment of active cultural industries making products that can be physically exported to customers all over the world.

Structures of Industrial Production

As we have seen at earlier stages in the argument, cultural industries of many different sorts are burgeoning in cities far and wide, and are in many cases now being actively fostered by policymakers in pursuit of local economic growth.

Several efforts have been made of late to assess the quantitative importance of cultural industries as a whole in various countries. Needless to say, such efforts are fraught with severe definitional problems. Even if a common definition of the cultural economy could be agreed upon, the disparate official industrial and occupational codes currently in use across the world would still make it impossible to establish fully comparable sets of accounts. All that being said, the published evidence, such as it is, suggests that cultural industries constitute an important and growing element of modern economic systems. Pratt (1997), for example, has shown for the case of Britain, that a little under 1 million workers (4.5 percent of the total labor force) are employed in cultural industries and their dependent sectors.[2] In another study, using a definition based on standard industrial categories, Scott (2000a) has indicated that cultural industries in the United States employ just over 3 million workers (2.4 percent of the total labor force). More importantly, these studies also indicate that employment in cultural industries is overwhelmingly located in large cities. Thus, Pratt's data show that London accounts for 26.9 percent of employment in British cultural industries. Scott's analysis indicates that in the United States just over 50 percent of all workers in cultural industries are concentrated in metropolitan areas with populations of 1 million or more, and of this percentage, the majority is actually to be found in just two centers, namely, New York and Los Angeles. Power (2002), following Pratt's definitional lead, finds that most workers in the Swedish cultural economy (which accounts for 9 percent of the country's total employment) are located in Stockholm. García et al. (2003) estimate that 4.5 percent of Spain's total GDP is generated by the cultural economy, with Madrid being by far the dominant geographic center.

Wherever they may occur, and like most other sectors that make up the new economy, cultural industries are typically composed of swarms of small firms often complemented by a more limited number of large establishments. Small producers in the cultural economy are much given to flexible specialization (Shapiro et al. 1992), or, in a more or less equivalent phrase, to neo-artisanal forms of production in which workers produce specialized outputs, such as television programs or clothing, but in short to medium series where design specifications are constantly changing (Eberts and Norcliffe 1998; Norcliffe and Eberts 1999). Big firms in the cultural economy occasionally produce in large volumes (which might tend to signify a diminution of symbolic function in final outputs), but are especially and perhaps increasingly prone to organization along the lines of "systems houses," a term used in the world of high-technology industry to indicate an establishment whose products are relatively small in number over any given period but where each individual unit of output represents huge inputs of capital and/or labor. The major Hollywood movie studios are classic cases of systems houses; other examples of the same phenomenon—or close relatives—are

---

[2] Much useful empirical information on the extent and diversity of cultural products industries in the UK can also be found in British Department of Culture Media and Sport (2001).

large publishers of magazines, major electronic games producers, television network operators, and, possibly, leading international fashion houses. Systems houses are of particular importance in the cultural economy because they so frequently act as the hubs of wider production alliances incorporating many smaller firms, and nowhere is this more the case than in the motion-picture industry of Hollywood (Scott 2005). Equally, they play a critical part in the financing and distribution of much independent production. We might add that large producers right across the cultural economy are increasingly subject to incorporation into the organizational structures of giant multinational conglomerates.

Despite the importance of large corporate entities, vertically and horizontally disintegrated networks of small specialized firms represent one of the more common modes of organizing production in the cultural economy. Such networks assume an assortment of guises, ranging from heterarchic webs of establishments to more hierarchical structures in which the work of groups of establishments is coordinated by a dominating central unit, with every possible variation between these two extreme cases. In short, as analysts like Caves (2000), Grabher (2004), Krätke (2002), Pratt (2000), Storper and Christopherson (1987) and others have repeatedly averred, much of the cultural economy can be described as conforming to a contractual and transactional model of production. The same model extends to the employment relation, with part-time, temporary, and freelance work being particularly prevalent. Within the firm, workers are often incorporated into project-oriented teams, a mode of work organization that is rapidly becoming the preferred means of managing internal divisions of labor in many of the more innovative segments of the modern cultural economy (Ekinsmyth 2002; Girard and Stark 2002; Grabher 2001; Heydebrand and Mirón 2002; Sydow and Staber 2002). By contrast, low-wage manual operators (e.g. in sectors such as clothing or furniture) are more apt to be caught up in piecework, often in sweatshop conditions.

To be sure, cultural industries characterized by transactional and labor-market situations like these almost always operate most effectively when the individual establishments that make them up exhibit at least some degree of locational agglomeration. This relationship reflects not only the economic efficiencies that so often flow from agglomeration as such but also the innovative energies that are unleashed from time to time in industrial clusters as information, opinions, cultural sensibilities, and so on are transmitted through them. As Molotch (1996, 2002) has argued, too, agglomerations of design-intensive industries acquire place-specific competitive advantages by reason of local cultural traditions and symbologies that become congealed in their products, and that imbue them with authentic character. These kinds of advantages correspond to what Chamberlin (1933) had in mind in his theory of monopolistic competition, just as Ricardo (1817) referred to the associated surplus profits as monopoly rent. A further ingredient of success for many kinds of cultural agglomerations is their irresistibility to talented individuals who flock in from every distant corner not only because they offer relevant forms of employment but also because these are the places where professional fulfillment can consistently be best pursued (Blau 1989; Montgomery and Robinson 1993). Menger (1993) refers to this phenomenon in terms of a dynamic of "artistic gravitation."

Place and Cultural Production

All of this suggests, again, that a tight interweaving of place and production system is one of the essential features of the new cultural economy of capitalism. In cultural industries, as never before, the apparatus of production and the wider urban and social environment merge together in potent synergistic combinations. Some of the most advanced expressions of this propensity can be observed in great world cities like New York, Los Angeles, Paris, London, or Tokyo. Certain districts in these cities are typified by a more or less organic continuity between their place-specific settings (as expressed in street scenes, shopping and entertainment facilities, and architectural landscape), their social and cultural infrastructures (museums, art galleries, theaters, and so on), and their industrial vocations (advertising, graphic design, audiovisual services, publishing, or fashion clothing, to mention only a few). Indeed, such cities often seek to promote this continuity by consciously reorganizing critical sections of their internal spaces like theme parks and movie sets, as exemplified by Times Square in New York, The Grove in Los Angeles, or the Potsdamer Platz in Berlin (Roost 1998; Zukin 1991). Soja (2000) has described projects like these under the rubric of "simcities," signifying the theatricalization of the built environment as a setting for everyday urban life and work. Hannigan (1998) uses the term "fantasy city" to allude to much the same phenomenon. In a city like Las Vegas, the urban environment, the production system, and the world of the consumer are today all so tightly interwoven as to form an indivisible unity (Gottdiener, Collins, and Dickens 1999). The city of work and the city of leisure increasingly interpenetrate one another in today's world.

The Local–Global Geography of Cultural Industries

Cultural-Products Industrial Districts

These trends are in significant ways rooted in the enormous expansion of cultural-products industries in contemporary cities, and they are given further expression in the upsurge of specialized cultural-products industrial districts in major urban areas. This upsurge has been captured in a burgeoning international literature that is continually bringing new and far-flung examples to light. Of the innumerable case studies that have been published over the last decade or so, the following may be advanced as a very small but representative sample: Arai et al. (2004) on internet business clusters in Tokyo; Bassett et al. (2002) on the making of nature films in Bristol; Calenge (2002) on the Parisian recorded music industry; Gibson (2002) on popular music in New South Wales; Hutton (2000) on design services in Vancouver; Krätke and Taylor (2004) on global media cities; Norcliffe and Rendace (2003) on comic book production in North America; Pollard (2004) on the jewelry industry of Birmingham; Rantisi (2002) on fashion design in New York; Scott (2000*b*, 2005) on both French cinema and Hollywood; and Yun (1999) on multimedia in Singapore. We can greatly extend the list if we include entertainment districts and urban "scenes" (Currid 2007; Lloyd 2002; Lloyd and Clark 2001; Montgomery 2007), together with cultural districts comprising museums, art galleries, and performing arts complexes (Brooks and Kushner 2001; Frost-Kumpf 1998; Santagata 2002; Van Aalst and Boogaarts 2002). A useful preliminary classification of all such districts has been drawn up by Santagata (2002).

In this section I propose to develop a more extended taxonomy that will help to shed additional light on the genesis and diversity of cultural-products industrial districts and their relation to urbanization. The suggested taxonomy is laid out in Figure 5.1. The details

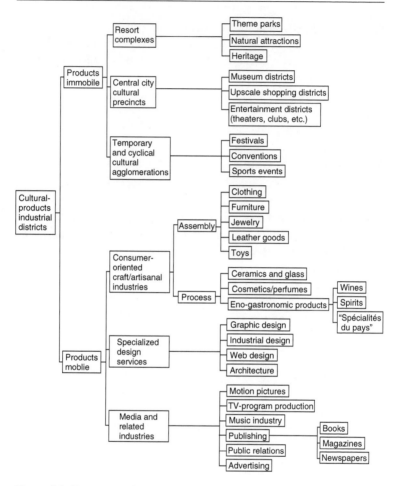

Figure 5.1. Taxonomy of cultural industrial districts

presented in the figure are not meant to be exhaustive, but only to suggest some of the more indicative features of these sorts of districts. Note that the categories given in any vertical slice of the diagram are far from always being mutually exclusive, and much overlap between them exists in reality. In Figure 5.1, an initial division of cultural industrial districts is made into those whose outputs are immobile and must therefore be consumed at the point of production (like tourist services), and those whose outputs are mobile and can be sold anywhere. In a very rough sort of way, these two divisions may be identified, respectively, with generation 1 and generation 2 policy approaches to local economic development as discussed earlier. The first division then decomposes into resort complexes, central-city cultural precincts, and temporary or cyclical cultural agglomerations. The second division is broken down into consumer-oriented craft and artisanal industries (themselves classified according to whether they are organized according to assembly or process methods of production), specialized design services, and media and related industries. These various branches of the taxonomy lead into a wide assortment of specific instances of cultural industries/districts, ranging from types based on theme parks and natural attractions, through clothing, furniture, and jewelry, to agglomerations of public relations and advertising firms. Finally, in the far left-hand vertical slice of Figure 5.1, I have indicated that even these detailed categories can be further unpacked, taking as examples (*a*) eno-gastronomic products, which are in turn divided into wines, spirits, and "spécialités du pays" (these being instances of cultural outputs with a strong agricultural connection), and (*b*) publishing which is represented by books, magazines, and newspapers. The types of industrial districts shown in Figure 5.1 all take on the guise of clusters of producers and associated local labor markets tied together in functional relations that generate complex economies of agglomeration. Even in the case of temporary and cyclical agglomerations where a single central facility may dominate the entire supply system, a dependent nexus of business activities and supporting services almost always develops in adjacent areas.

Individual metropolitan areas, of course, are commonly endowed with many different classes of cultural districts. Los Angeles is a dramatic illustration of this point, with its numerous clusters (some of which are shown in Figure 4.1) based on industries such as clothing, furniture, jewelry, motion pictures, television-program production, music recording, publishing, and advertising, as well as its array of theme parks, convention centers, and sports facilities, and its upscale shopping and entertainment districts (Jencks 1993; Molotch 1996; Scott 2000*a*). The Los Angeles metropolitan area also contains a cluster of highly reputed architectural firms and is the site of what is probably the world's largest collection of automobile design studios. Moreover, the Los Angeles example illustrates the point that positive spillover effects frequently diffuse across the entire urban area from their more narrowly confined district of origin. Thus, design practices or fashion innovations that appear in one district are often imitated in others; similarly, a particular district in the city may generate specific kinds of worker skills and sensibilities that are then found to have critical applications in other parts of the same city; and reputation effects that accrue to a particular industry (e.g. motion pictures) in a particular place can sometimes be appropriated by other industries (e.g. fashion clothing) in adjacent locations. In Los Angeles, the latter relationship has been notably consolidated in recent years in the annual Oscar ceremony marked as it is by a highly mediatized combination of film personalities and fashion displays.

The Gobal Connection

Despite the predisposition of firms in cultural industries to cluster in close mutual proximity to one another, their outputs in many cases flow with relative ease across national borders and are a steadily rising component of international trade. As new web-based distribution technologies are perfected, this process of globalization will assuredly accelerate, at least where digitization of final products is feasible (Currah 2003). Observe that globalization does not necessarily lead to the locational dispersal of production itself. On the contrary, globalization *qua* spatial fluidity of final products often helps to accentuate agglomeration because it leads to rising exports combined with expansion of localized competitive advantages (see Chapter 7). Concomitant widening and deepening of the social division of labor at the point of production then intensifies agglomeration, especially where intra-agglomeration transactions costs remain high. Locational agglomeration and globalization, in short, are complementary processes under specifiable social and economic circumstances. That said, the falling external transactions costs associated with globalization do sometimes undermine agglomeration from the other end, as it were, by making it feasible for some kinds of functionally embedded production activities to move to alternative locations. It is now increasingly possible for certain types of functions that could not previously escape the centripetal force of agglomeration to decentralize to alternative locations where cost conditions (especially labor costs) are relatively attractive. This locational syndrome is above all characteristic of sectors or subsectors that are relatively streamlined in regard to their basic transactional relations and labor requirements, such as plants processing CD-ROMs for the recording industry, or call centers in the communications business, or film shooting activities in the motion-picture industry (i.e. runaway production). Sometimes this sort of decentralization leads to the formation of alternative clusters or satellite production locations, as illustrated by the sound stages and associated facilities that have come into existence in Toronto, Vancouver, Sydney, and other parts of the world in order to serve the film-production companies of Hollywood (Coe 2000; Goldsmith and O'Regan 2005; Scott and Pope 2007).

The overall outcome of these competing spatial tensions in the modern cultural economy is a widening global constellation of centers producing many different outputs. The logic of agglomeration and increasing-returns effects suggests, *ceteris paribus*, that ultimately—in a world of declining transportation costs—one globally dominant center should emerge in any given sector, at least for industries where all outputs, no matter what their point of origin, are perfect substitutes for one another. But in a world economy that is moving ever more decisively toward monopolistic competition, this ultimate possibility is actually becoming less and less likely. Even in the case of the motion-picture industry, which is currently dominated worldwide, and to a massive degree, by Hollywood, it can be plausibly argued on the basis of place-based product differentiation processes that multiple production centers will continue to exist if not to flourish. Large multinational corporations play a decisive role across this entire functional and spatial field of economic activity, both in coordinating local production networks and in operating worldwide distribution and marketing systems (Hoskins, McFadyen, and Finn 1997; Nachum and Keeble 2000). In the past, multinationals based in the United States have led the race to command global markets for cultural products of all categories, but firms from other countries are now entering the fray in increasing numbers, even in sectors like the media that have hitherto been considered as the privileged preserve of North American firms (Herman and McChesney 1997; Krätke and Taylor 2004). This circumstance again enhances the likelihood that global cultural production will become increasingly diversified in the future.

The opening up of global trade in cultural products, then, does not so much appear to be resulting in the emergence of just a few hegemonic centers in any give sector, but on the contrary is helping many different production centers around the world to establish or reestablish durable competitive advantages and to attack new markets. The film and television industries illustrate this point with some force. To begin with, there is something like an irreducible corpus of television-program production activities in the vast majority of countries, if only because of the persistent preferences that most societies display for local content in top-rated TV programs. This means that a basic domestic production capacity will almost certainly continue to thrive at least in larger countries, thus also providing a foundation for competitive forays into new markets and products. Several countries, too, especially in Asia and Western Europe, retain sizable motion-picture industries, and in some cases these are showing new signs of life (see Figure 5.2). Certain national production centers like Bangkok, Beijing, Bombay, Hong Kong, Seoul, Tokyo, Cairo, Teheran, Berlin, London, Madrid, Paris, and Rome have never fully capitulated to Hollywood, even though most of them have suffered severe competitive depredations at various times in the past. Several of these centers have produced films that have performed very successfully on selected markets in recent years, and a number of them today are clearly girding up for a new round of market contestation with Hollywood. While the commercial supremacy of the American film industry is unlikely to be broken at any time in the foreseeable future, at least some of these other centers will conceivably carve out stable niches for themselves in world markets, and all the more so as they develop more effective marketing and distribution capacities, and as homegrown media corporations acquire increasing global muscle. Bollywood's recent attempts at outreach to world markets are a symptom of this trend. The international successes of Hong Kong action films, *anime* cartoons made in Tokyo, wide-canvas dramatic features from Beijing, or Latin American telenovelas, all suggest a similar conclusion. This argument, if correct, points toward very much more diversity across the global audiovisual production system in the future than was thought possible by many analysts in the past, especially those espousing the blunt cultural imperialism argument (e.g. Mattelart 1976; Michalet 1987).

At the same time, different cultural industrial agglomerations around the world are increasingly caught up with one another in global webs of coproductions, joint ventures, creative partnerships, and so on. In this manner, widely dispersed productive combinations are surging to the fore, drawing on the specific competitive

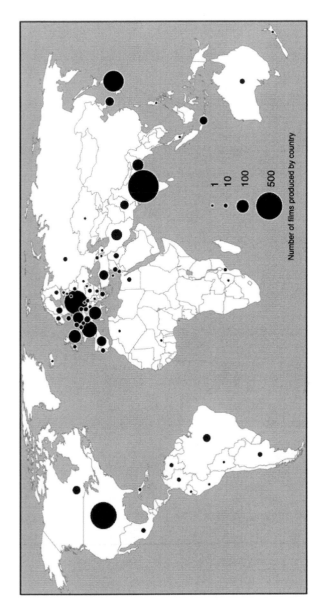

Figure 5.2. Number of long films produced by country. Data are for the most recent available year over the period 1995–9

Source: UNESCO, http://stats.uis.unesco.org/TableViewer/tableView.aspx?ReportId= 12. NB: No data are given for China or Hong Kong.

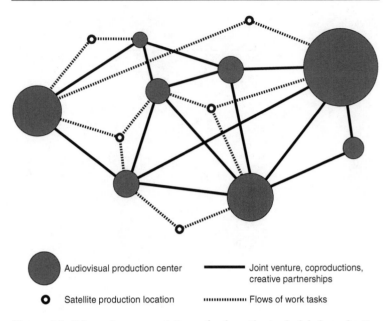

Figure 5.3. Schematic representation of a hypothesized global production landscape in the audiovisual industries

advantages of diverse clusters without necessarily compromising the underlying force of agglomeration itself. Thus, many new media firms in San Francisco work intensively with Hollywood film-production companies, and have also been observed to establish partnerships with book publishers in New York and London (Scott 1998a). Film stars from places like Bombay, Hong Kong, and Tokyo sell their place- and culture-specific human capital to production companies in North America and Europe, a practice that in turn enhances the market power of the films they make when they return to their home base (Yau 2001). Pathania-Jain (2001) has shown that multinational media corporations like BMG, Disney, EMI, News Corp., Polygram, and Sony are currently building collaborative alliances with Indian firms not only in order to penetrate Indian markets with their own products but also to tap into the productive and creative capacities of Mumbai's Bollywood. The general point can be summarized by reference to Figure 5.3 which is meant to represent a hypothetical landscape of the audiovisual industries at some time in the not-too-distant future. Figure 5.3 portrays in essence a landscape of global extent punctuated by occasional dense production agglomerations and an associated system of satellite centers, some of which may eventually mature into full-blown agglomerations in their own right. This example is perhaps especially provocative, and it is presented here to suggest, contrary to much received opinion, not only that a multiplicity of audiovisual production centers will in all probability continue to function in the future but also that numbers of them can be expected to prosper and grow.

It need scarcely be pointed out that this scenario is highly speculative, and things may well fall out otherwise, depending on a hundred different unforeseen contingencies. Despite this warning, it seems that several kinds of cultural sectors besides the motion-picture industry (music, architectural services, or publishing, for example) are also poised at the brink of analogous developments. In all cases, the projected evolutionary trend forward involves a diverse array of production agglomerations spatially distributed around the world, each commanding distinctive market segments, even where one particular center dominates overall. Where appropriate local assets are available and effectively mobilized in the service of niche markets even quite small centers can maintain a lasting presence in the global cultural economy.

Culture and Local Economic Development Policy

The burning question that crops up once again at this point is in what ways, if at all, can urban and regional policymakers take advantage of complex trends like these for the purposes of local economic development? There are good reasons, obviously, for presuming that cultural industries are, or ought to be, of compelling interest to policymakers. As we have seen, these industries are growing rapidly; they tend (though not always) to be environmentally friendly; and they frequently (though again not always) employ high-skill, high-wage, creative workers. Equally, cultural industries usually contribute significantly to the quality of life in the places where they congregate and enhance the image and prestige of the local area. However, they cannot be straightforwardly conjured into existence by simple acts of political will or fiscal prodigality. Just as local governmental authorities all over the United States threw huge sums of money out of the window in the 1980s and 1990s in the quest to build new high-technology industrial clusters, so can we predict parallel miscarriages of policy in years to come as efforts to build various new Hollywoods or the next Silicon Alley materialize. Careful and theoretically informed assessments of available opportunities and inherent constraints are essential if such miscarriages are to be avoided, and we must always be prepared for the possible conclusion in any given case that the best course of action is in fact to do nothing (Cornford and Robins 1992).

Urban and regional economic development agendas focused on cognitive-cultural sectors at large need to be especially clear about the internal dynamics of the dense agglomerations that are one of the primary expressions of these sectors' geographic logic. The essential first task that policymakers must face—above and beyond a simple audit of the resources and capacities already in place—is to assess the actual and potential collective order of the local economy along with the multiple sources of the increasing-returns effects that invariably crisscross through it. It is this collective order more than anything else that presents possibilities for meaningful and effective policy intervention. Blunt top-down approaches focused on indicative planning are rarely likely in and of themselves to accomplish much at the local scale, except in special circumstances. In terms of cost–benefit ratios and general workability, the most successful types of policies will as a rule be those that concentrate on promoting detailed external economies of scale and scope. The point here is both to stimulate the formation of useful agglomeration effects that would otherwise be undersupplied or dissipated in the local economy and to ensure that existing external economies are not subject to severe misallocation. Finely tuned bottom-up measures are essential in situations like this. Specifically, policymakers need to pay special attention to promoting (*a*) collaborative interfirm relations and reciprocity in order to mobilize latent synergies, (*b*) efficient, high-skill local labor markets, and (*c*) local sources of industrial creativity and innovation. Certainly the practical details involved in the implementation of policy measures directed to these ends will frequently be extremely difficult to deal with depending on a host of different local circumstances. However, institution-building in order to promote cooperative effort among different groups of local actors must surely be seen as one basic condition of success. Complementary lines of attack involve approaches such as the initiation of labor-training programs, the setting up of centers for the encouragement of technological upgrading or design excellence, organizing exhibitions and export drives, and so on, as well as socio-juridical interventions like dealing with threats to the reputation of local product quality due to free rider problems, or helping to protect communal intellectual property. In addition, an appropriately structured regional joint private–public partnership could conceivably function in many instances as a vehicle for generating early warning signals as and when the local economy appears to be in danger of locking into low-level equilibrium due to adverse path selection dynamics (Cooke and Morgan 1998; Storper and Scott 1995). The latter problem is prone to make its appearance in localized production systems where complex, structured interdependencies often create long-run developmental rigidities.

In practice, and notwithstanding these broad illustrative guidelines, there can be no standardized or boilerplate approach to the problem of local economic development. Each case needs to be treated on its own merits, paying full attention to the historical and geographical idiosyncrasies that characterize every individual place. This admonition is doubly emphatic in regard to the cultural economy, given the enormous heterogeneity of its production activities, its high dependence on workers' personalized attributes, and its sensitivity to subtle place-specific forces. A simple but sound precept guiding any plan of action in regard to urban and regional economic development, no matter what specific sectors may be at stake, is to start off with what already exists, and to build future expectations around whatever concrete opportunities this initial position seems to offer. Cities or regions that lack any preexisting base of cultural production face a more refractory policy problem. Yet even where no obvious prior resources are available, it has occasionally been feasible to initiate new pathways of development based on cultural industries. Recall the examples of the old industrial areas cited earlier. Much new development in these areas has focused on building a new cultural economy by means of conscious efforts to use the relics of the industrial past as core elements of a reprogrammed landscape of production and consumption. Local authorities right across Europe and North America are striving to revalorize inner city areas on the basis of experiments like these, often in concert with local real estate interests.

A particularly striking case of a local economic development project that has attempted to conjure up a major new focus of urban and cultural economic activity virtually *ex nihilo* is presented by the Multimedia Super Corridor project in Malaysia (Bunnell 2002*a*, 2002*b*; Indergaard 2003), a project that was initiated on the basis of little more than political will and an ambitious faith in large-scale top-down planning. Work on the project started in 1996, but was immediately put in jeopardy by the Asian fiscal crisis of 1997–8. The Super Corridor remains nonetheless a priority of the Malaysian government, and its physical development continues to move forward. The project represents a massive infrastructure and urbanization effort stretching 30 miles southward from Kuala Lumpur to the new international airport. It comprises two main functional centers, Putrajaya, where governmental administrative offices are being increasingly concentrated, and Cyberjaya, which is projected to develop as a major cluster of software, information, and new media producers. The scheme is being planned in large degree as a pivot of new economic and cultural growth in Malaysia, taking particular advantage of the country's strategic location at the center of an immense potential market of Chinese, Arabic, and Indian consumers. The Malaysian workforce, moreover, embodies all the linguistic skills and cultural sensibilities required to deal with this market on its own terms. Needless to say, the Multimedia Super Corridor project is fraught with severe risks. It will no doubt eventually bear fruit of some sort, but whether the long-term benefits will outweigh the enormous costs remains very much an open question at this stage.

Finally, policymakers have to keep a clear eye on the fact that any industrial agglomeration is dependent not only on the proper functioning of its complex internal relationships but also on its ability to reach out to consumers in the wider world. Successful agglomerations, in short, must always be possessed of adequate systems for marketing and distributing their outputs. This matter is of special importance with respect to cultural products because they are subject above all to symbolic rather than utilitarian criteria of consumer evaluation, and in many cases are dependent on peculiar kinds of infrastructures and organizations for their transmission. In a situation of intensifying global competition, effective distribution is critical to survival and indispensable for growth (Greffe 2002; Scott 2000*b*). It might be contended, for example, that the poor commercial performance of French films in export markets is not so much due to linguistic barriers—and certainly not to a lack of basic production capacity or talent—as it is to the competitive deficiencies of French film marketing and distribution systems outside of France. Partial redress of these deficiencies might be secured through a shift in policy by the Centre National de la Cinématographie (the central government–industry body responsible for oversight of the French cinema) toward lower levels of subsidized production and higher levels of subsidized distribution. A clear recognition of the general importance of distribution is expressed in the European Union's Media Plus Program initiated in January 2001 in succession to the earlier Media I and Media II programs. A principal objective of the Program is to strengthen international distribution systems for European audiovisual products. Concerted efforts by policymakers in cultural agglomerations all over the world to upgrade marketing and distribution systems for local outputs are surely one of the fundamental keys to effective participation in the hypothetical global production landscape as projected in Figure 5.3.

The Road Ahead?

The notion of the cultural economy as a source of local economic development is still something of a novelty, and much further reflection is required if we are to understand and exploit its full potential while simultaneously maintaining a clear grasp of its practical limitations. In spite of these caveats, the cultural economy continues to expand apace in large metropolitan areas, as well as in selected smaller centers, and as it does so, it is opening up new opportunities for policymakers to raise local levels of income, employment, and social well-being. While most development based on cultural industries will in all likelihood continue to occur in the world's richest countries, a number of low- and middle-income countries, such as Brazil, China, and India, are finding that they too are able to participate in various ways in the new cultural economy and even to sell significant volumes of output on global markets. Even old and economically depressed industrial areas can occasionally turn their fortunes around by means of well-planned cultural initiatives. The great global city-regions of the advanced capitalist countries represent in practice the high-water marks of the modern cultural economy. This proposition refers not only to the many and diverse individual sectors of cultural production that are usually located in these cities but also to their wider environmental characteristics and global connections. Some sections of major city-regions today display a remarkable systemic unity running from the physical urban tissue, through the cultural-production system as such, to the very texture of local social life. These features, indeed, are mutually constitutive elements of much of the contemporary urbanization process. Central Paris with its monumental architectural set pieces, its museums and art galleries, its intimate forms of street life, and its traditional artisanal and fashion-oriented industries represents a notably symbiotic convergence of built form, economy, and culture. Even in well-established cases like this, policymakers still have a major role to play by intervening at critical junctures in the production system and the urban milieu in order to release synergies leading to superior levels of product appeal, innovativeness, and competitiveness.

As a corollary, we seem to be moving steadily into a world that is becoming more and more cosmopolitan and eclectic in its modes of cultural consumption. Certainly for consumers in the more economically advanced parts of the world, the standard American staples are now but one element of an ever-widening palette of cultural offerings comprising Latin American telenovelas, Japanese comic books, Hong Kong kung fu movies, West African music, London fashions, Balinese tourist resorts, Australian and Chilean wines, Mexican cuisine, Italian furniture, and untold other exotic fare. This trend is in significant degree both an outcome of and a contributing factor to the recent, if still in many ways incipient advent of an extensive global system of cultural agglomerations. All the same, the global reach of many cultural-products agglomerations, magnified as it is by the commercial prowess of the multinational corporations with which they are so frequently associated, has not always been attended by benign results. This situation has in fact led to numerous political collisions over issues of trade and culture. One of the more outstanding instances of this propensity is the clash that occurred between the United States and Europe over trade in audiovisual products at the time of the GATT (now WTO) negotiations in 1993. A further critical set of concerns revolves around issues of cultural politics generally, not only in regard to development and trade but also and perhaps yet more significantly, in regard to matters of human growth and progress at large. As cryptic as this remark may be, it opens up a vast terrain of debate about the qualitative meaning of the overarching system of cultural consumption that is being ushered into existence by the trends and processes discussed above. The goods and services that sustain this system are to ever-increasing degrees fabricated within production networks organized according to general principles of capitalist enterprise, even if they are increasingly found within far-flung industrial clusters. This is a world, as Lash and Urry (1994) have written, in which culture is produced overwhelmingly in commodity form, while commodity production itself becomes more and more deeply infused with aesthetic and semiotic meaning. One important effect of this condition is the increasing diversity of cultural products across the world; but another is their pervasive ephemerality and waning symbolic intensity (Jameson 1992). A vibrant cultural politics attuned to these issues should no doubt attempt to intensify the push to diversity while seeking to mobilize opinion in favor of a global cultural economy that promotes intelligence and sensibility rather than their opposites. Even in the era of the cognitive-cultural economy, an informed, critical, and contentious body of consumers is in the end the most robust bulwark keeping the social dysfunctionalities predicted by the Frankfurt School at bay.

6

Chiaroscuro: Social and Political Components of the Urban Process

Introduction

Cities in capitalism have always developed first and foremost in relationship to the pressures of production and exchange. But they are also critical foci of social and political life, and their character in these regards reconfigures their purely economic role in many and important ways. Indeed, the city, as such, only emerges in its full form when the social and the political aspects of communal existence are each in place and integrated in complex reflexive relations with the economic order. In the absence of mechanisms of production and exchange, the livelihood of every urban denizen is in jeopardy; in the absence of viable processes of social reproduction, the economy of the city must necessarily wither away; and in the absence of some forum of political consultation and decision-making, the endemic internal breakdowns and collisions of interest in the city will steadily tend to spin out of control. The latter point may be highlighted by noting that the political dynamics of any given city are inevitably subject to many different frictions flowing from the structural failures, the inequalities, and the negative lock-in effects that come constantly to the fore in the urban environment. These frictions may in certain instances be adjudicated by institutions at the national level, but since they derive specifically from maladjustments in the urban sphere, they will often most effectively be handled within the framework of city governance as such.

More generally, economic and social inequalities together with their attendant conflicts have always been a feature of cities throughout the history of capitalism, though their precise shape and form vary widely depending on the larger historical context. In the classical nineteenth-century factory town, they were expressed above all in the sharp class division between the property- and capital-owning bourgeoisie on the one hand and the mass of impecunious proletarian workers on the other. In the fordist metropolis of the mid-twentieth century, they were primarily evident in the split that separated a white-collar fraction domiciled above all in the suburbs from a blue-collar fraction that was mainly confined to inner city areas marked in significant degree by inferior housing, social breakdown, and ghettoization along racial and ethnic lines. In the post-fordist world that has emerged over the last couple of decades, an alternative bipartite intra-urban division has appeared. Here, a basic (and deepening) line of socioeconomic cleavage runs between an urban elite that participates actively in the high-wage, high-skill employment opportunities offered by the new cognitive-cultural economy, and an expanding body composed of the working poor (many of them immigrants), the unemployed, and the homeless. In many respects, the prospects of those at the very bottom of the socioeconomic ladder today remain more or less permanently limited. Notwithstanding these different kinds of inequalities, the contrasting social worlds that have existed in large cities throughout the history of capitalism never represent functionally separate universes; on the contrary, they are always organically intertwined with one another through the peculiar employment regimes and divisions of labor that characterize each phase of urban economic development. In today's cognitive-cultural economy, the conspicuous split between the upper and lower fractions of the labor force has been accentuated by the erosion of stable middle-level jobs from the American economy as a result of technological change and the shift of large segments of economic activity to low-cost locations in other countries. This state of affairs inevitably colors all aspects of social and political life in the contemporary metropolis.

Affluence and Poverty in American Cities

In the immediate post-War decades, disparities in income and life chances were to a degree held in check in large American cities by a series of social and institutional arrangements that maintained a floor under the wages of the working class and that effectively redistributed income from higher to lower social strata. The widespread pattern of employment of blue-collar workers within the framework of the large industrial corporation, with its generous wage rates and its complex machinery of personnel management, was itself a significant factor in helping to maintain overall living standards. The passage of the Wagner Act of 1935, which removed impediments to unionization and established a framework for the orderly exercise of collective bargaining rights by the workforce at large, helped further to consolidate the relative prosperity of the working class in American cities. The full-blown welfare statist policy apparatus that was put into effect in the decades following World War II gave yet more stability and order to the lives of workers by reason of its many different programs of public spending directed to education, health, social services, housing, and so on, and its generous provisions for unemployment insurance.

The crisis of the mass production system and its associated policy arrangements in the mid- to late 1970s severely undercut this model of social development, and ushered in a period of significant economic hardship accompanied by unusually high levels of job loss and fiscal stress, most especially in the large metropolitan areas of the American Manufacturing Belt and the equivalent industrial zones of Western Europe. The election of Ronald Reagan as president of the United States in 1980 (and of Mrs Thatcher as prime minister of the United Kingdom in 1978) sounded the death-knell of keynesian-welfare statism in its classical form and marked a series of shifts toward the more neoliberal policy stance that has since spread steadily throughout the advanced capitalist world. Concomitantly, a robust post-fordist economy was now starting to gather momentum, and as it has further developed over the subsequent decades, it has come more and more to assume the cognitive-cultural configuration as described in earlier chapters of this book. As I have argued, much of the recent shift observable in the overall economic structure of the advanced capitalist societies must be put in the context of an underlying historical transition from the second to the third machine ages, that is, away from the rigid large-scale mechanical technologies of mass production to flexible digital technologies and computerization. One of the major consequences of this shift has been the steady—though still far from complete—replacement of routinized labor processes (both mental and manual) in the core capitalist economies by forms of work that call much more insistently for exercise of the discretionary cognitive and cultural capacities of the labor force. To be sure, there remain sectors in the advanced capitalist countries where resistance to this shift is in evidence, but as shown in Chapter 3, sectors like these are increasingly relocating to places relatively low down in the urban hierarchy, and also, in ever great degree, to cheap labor depots in the world periphery. The net consequence is that the cognitive-cultural economy is becoming more and more sharply dominant in major metropolitan areas today.

In contrast to the pervasive deskilling that occurred in the fordist period, much recent technical change appears generally to be skills-enhancing, and this state of affairs is no doubt one of the main factors underlying the widening gap between the incomes of more and less skilled workers in large cities. This statement might perhaps be qualified by noting that the gap can also be described in terms of a division between workers who hold formal qualifications and those who do not. Even at the bottom end of the labor market, workers are not always devoid of skills, far from it, for they are increasingly engaged in relatively flexible work activities that require them to exhibit significant degrees of practical know-how and interpersonal sensitivity, and this is especially the case in regard to work in large metropolitan areas. The deep contrasts in wages and living standards that characterize the new economy in the United States have been amplified by the changes in the overall institutional environment that have been going on over the last couple of decades, above all in the context of a disappearing welfare state and the deepening of global competitive pressures. In parallel with these changes, a number of critical shifts in the structure of labor markets and the employment relation have also contributed to mounting levels of worker inequality, as manifest notably in the declining power of workers' organizations, the massive feminization of the labor force, and the ever-extending sweep of flexible work and labor-market arrangements.

Some important urban implications of these remarks are highlighted in the data laid out in Figure 6.1. This figure displays census information on individual wage and salary incomes across all metropolitan areas in the United States as a whole for five different years over the period from 1970 to 2005. Wages and salaries rather than family income data are the preferred unit of measure here since they directly reflect returns to work. However, even when the

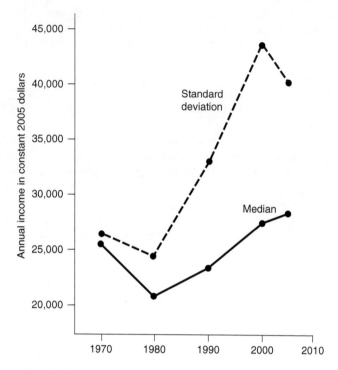

Figure 6.1. Median and standard deviation of individual wage and salary incomes in all US metropolitan areas

Source: Decennial Censuses and American Community Survey, accessed through IPUMS USA, Minnesota Population Center, at http://usa.ipums.org/usa/.

information given in Figure 6.1 is reexpressed on the basis of family incomes, the same basic patterns emerge. Recall that the number of metropolitan areas as well as their spatial definitions tends to change from one census to the next and no attempt has been undertaken to make corresponding adjustments in the data presented in this figure. The net effect of this decision is undoubtedly to introduce some bias into the analysis, but the degree of distortion is probably not great for the new metropolitan areas that crop up in any given census tend to lie toward the bottom end of the urban hierarchy (so that dominant patterns remain relatively undisturbed) and enlargement of the boundaries of existing metropolitan areas occurs almost entirely at the extensive margins of urbanization (where population densities

Table 6.1. Percentage frequency distributions of wage and salary incomes by metropolitan size category, 2005

Wage and salary level	Metropolitan size category[a]				
	1	2	3	4	5
<25,000	40.7	42.4	46.9	47.2	50.5
25,000–50,000	29.3	31.3	31.9	31.4	31.4
50,000–75,000	15.5	14.7	12.9	12.8	11.5
75,000–100,000	6.6	6.0	4.2	4.2	3.4
100,000–125,000	3.2	2.4	0.9	1.2	0.9
125,000–150,000	0.7	0.3	0.4	0.5	0.2
>150,000	4.0	2.9	2.8	2.7	2.1
Total	100	100	100	100	100

[a]Metropolitan size categories are defined in terms of population levels in the year 2000, that is (1) 5,000,000 and above, (2) 1,000,000–5,000,000, (3) 500,000–1,000,000, (4) 250,000–500,000, and (5) 250,000 and below.

Source: American Community Survey, accessed through IPUMS USA, Minnesota Population Center, at http://usa.ipums.org/usa/.

are relatively low). As Figure 6.1 clearly reveals, median wages and salaries (in constant 2005 dollars) in American metropolitan areas declined over the crisis years of the 1970s, but have subsequently risen steadily since then.[1] However, the figure also demonstrates that the dispersion of wages and salaries has increased at an even more rapid rate since 1980, though a small reversal of this trend can be observed for the period from 2000 to 2005. In other words, real wage and salary levels in American metropolitan areas have tended to rise in recent decades, but inequalities have increased even more markedly.

Now let us consider the situation where the wages and salary data are broken down by metropolitan size category. The five main categories already established in Chapter 3 are deployed for this purpose. Table 6.1 lays out frequency distributions for all five metropolitan size categories showing the percentage of employees in different wage and salary brackets in the year 2005. Three main points can be made on the basis of a scrutiny of Table 6.1. First, all five frequency distributions are obviously heavily skewed to lower wage and salary levels, and in each case, as many as 40–50 percent of employees are

[1] Note that these and subsequent calculations involving wage and salary statistics are based on micro-data taken from the census. Cases where a zero wage or salary level is reported are eliminated from the analysis.

Table 6.2. Median wage and salary incomes by metropolitan size category, 1970–2005; all values in constant 2005 dollars

Year	Metropolitan size category[a]					All metropolitan areas	Total number of metropolitan areas
	1	2	3	4	5		
1970	28,943	25,419	24,915	25,419	27,900	25,400	119
1980	24,375	23,713	21,343	20,964	20,063	21,341	256
1990	29,885	26,897	23,524	22,414	20,920	23,968	249
2000	30,621	29,488	26,085	25,972	22,683	27,220	283
2005	30,560	30,400	26,845	26,482	24,448	27,504	283

[a]Metropolitan size categories are defined in terms of population levels in the year 2000, that is (1) 5,000,000 and above, (2) 1,000,000 to 5,000,000, (3) 500,000 to 1,000,000, (4) 250,000 to 500,000, and (5) 250,000 and below.

Source: Decennial Censuses and American Community Survey, accessed through IPUMS USA, Minnesota Population Center, at http://usa.ipums.org/usa/.

in the lowest category represented by incomes of $25,000 or less per annum. Second, wages and salaries tend on average to be higher for larger metropolitan size categories than for smaller. No doubt we can account for this broad difference in significant degree by invoking both the presumed higher levels of productivity and the higher cost of living associated with large-scale urban agglomeration (Glaeser and Maré 2001; Simon and Nardinelli 2002). Third, the right-hand tails of the distributions are longer for larger metropolitan size categories than for smaller, signifying the relatively greater presence of a super-elite of workers in big cities.

The information shown in Tables 6.2 and 6.3 provides further elaboration on these comments. These tables offer data on median wage and salary levels and on the standard deviations of wage and salary incomes for each metropolitan size category for the period from 1970 to 2005. Table 6.2 indicates, again, that median wage and salary levels tended to decline over the 1970s, but then increased steadily up to 2005 for each metropolitan size category (apart from a very small setback for category 1 between 2000 and 2005). We also see clearly once more the general increase in wage and salary incomes as we move from smaller to larger metropolitan areas. Table 6.3 for its part reveals the striking increase in income inequality that occurred from 1980 to 2000, followed by a small reversal after 2000. Moreover, inequalities are very much greater in larger than in smaller metropolitan areas, and the contrast has become yet more pronounced with the passage

Table 6.3. Standard deviations of wage and salary incomes by metropolitan size category, 1970–2005; all values in constant 2005 dollars

Year	Metropolitan size category[a]					All metropolitan areas	Total number of metropolitan areas
	1	2	3	4	5		
1970	30,140	28,141	25,607	26,807	24,350	26,511	119
1980	27,176	26,019	23,590	23,814	23,057	24,638	256
1990	39,315	34,486	31,023	30,463	28,115	32,678	249
2000	53,133	46,279	39,297	41,347	36,799	43,283	283
2005	49,644	42,482	35,593	38,469	32,812	40,000	283

[a]Metropolitan size categories are defined in terms of population levels in the year 2000, that is (1) 5,000,000 and above, (2) 1,000,000 to 5,000,000, (3) 500,000 to 1,000,000, (4) 250,000 to 500,000, and (5) 250,000 and below.

Source: Decennial Censuses and American Community Survey, accessed through IPUMS USA, Minnesota Population Center, at http://usa.ipums.org/usa/.

of time. In 1980, the standard deviation for the largest metropolitan size category was 1.18 times greater than that for the smallest; by 2005, it was 1.51 times larger.

The specific cases of the three largest metropolitan areas in the United States, namely, the New York–Northern New Jersey–Long Island MSA, the Los Angeles–Long Beach–Santa Ana MSA, and the Chicago–Naperville–Joliet MSA, shed much useful additional light on these variations in urban wage and salary incomes. Figure 6.2 shows frequency distributions of wage and salary incomes for these three metropolitan areas in the year 2005. The modes of the distributions obviously coincide with extremely low income levels. Observe that there is also a subsidiary peak in all three cases reflecting the truncation of the original statistics at the income level represented by $200,000 and above. For New York, Los Angeles, and Chicago, the standard deviations of wage and salary incomes are 46,298, 57,151, and 45,931, respectively, signifying unusually high levels of income inequality, especially in Los Angeles with its notably high proportion of immigrant workers. In addition, in each instance, the ratio of wage and salary levels at the 20th and 80th percentiles is of the order of one to five. At the 10th and 90th percentiles, the ratio is 1:17 and more.

These trends concern income disparities for the population as a whole, irrespective of demographic group. We must now take into account the fact that incomes also vary systematically in relationship

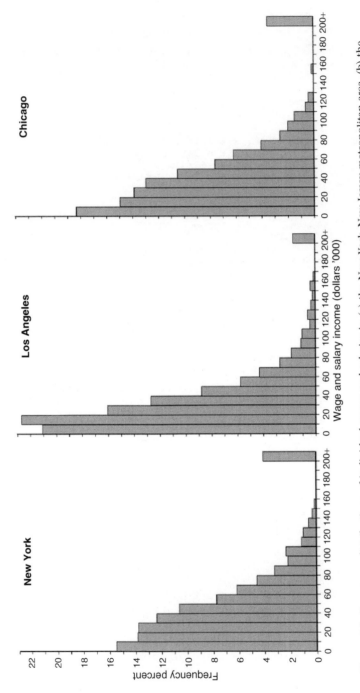

Figure 6.2. Frequency distributions of individual wages and salaries in (a) the New York–New Jersey metropolitan area, (b) the Los Angeles–Long Beach–Santa Ana metropolitan area, and (c) the Chicago–Naperville–Joliet metropolitan area, 2005

Source: American Community Survey, accessed through IPUMS USA, Minnesota Population Center, at http://usa.ipums.org/usa/.

Table 6.4. Percentage of different racial and ethnic groups living at or below the official poverty line in the New York–Northern New Jersey–Long Island MSA, the Los Angeles–Long Beach–Santa Ana MSA, and the Chicago–Naperville–Joliet MSA, 2005

	Percent of each racial or ethnic group at or below the poverty line		
	New York–Northern New Jersey–Long Island	Los Angeles–Long Beach–Santa Ana	Chicago–Naperville–Joliet
White	8.5	12.5	6.9
African-American	20.0	20.4	28.4
Hispanic origin	23.7	22.1	18.2
Asian	13.3	12.0	7.8

Source: American Community Survey, accessed through IPUMS USA, Minnesota Population Center, at http://usa.ipums.org/usa/. NB: White and Hispanic-origin populations are not necessarily mutually exclusive. (MSA = metropolitan statistical area.)

to different racial and ethnic categories. Table 6.4 presents data on poverty rates in the New York–Northern New Jersey–Long Island MSA, the Los Angeles–Long Beach–Santa Ana MSA, and the Chicago–Naperville-Joliet Chicago MSA by selected racial and ethnic categories in 2005. The official census definition of the poverty line in 2005 for a family of one was $10,160 per annum, and for a family of three it was $15,577. Table 6.4 shows that whites in the three metropolitan areas are in general much less likely than any other group to be living at or below the poverty level, followed by Asians, who, in the case of Los Angeles, actually have a slightly lower poverty rate than whites. By contrast, poverty rates among the African-American and Hispanic-origin populations are at least double and occasionally triple the rate for whites, and these two groups represent by far the most disadvantaged elements in the contemporary metropolis. Large numbers of individuals and families in both of these groups suffer from a stubborn syndrome that revolves around their relatively low educational levels, their social marginalization, and the prejudices of the wider host society. That said, the relations of African-American and Hispanic-origin populations to the wider urban economy seem to be evolving in rather different directions from one another. Hispanic immigrants (together with smaller complements of Asian immigrants) are integral to the operation of the lower reaches of economy in major cities, especially in activities such as electronics assembly, clothing and furniture manufacture, restaurant work, and so on. African-Americans, notably those who lack a high-school education, have not been absorbed to anything like the same extent into this segment of the urban economy, with the consequence that unemployment rates for this group are now exceptionally high in major US cities. There is some evidence, in fact, to suggest that a process of crowding out of African-American workers from the bottom end of labor market is occurring in cities as a consequence of preferences on the part of employers for yet more socially and politically marginalized workers, namely, immigrants, and above all recent immigrants from poor countries who can be relied upon not only to accept low wages but also to remain relatively passive in the face of generally substandard working conditions (Lichter and Oliver 2000; Scott 1996*b*; Waldinger 2001). In many large cities, the combined effects of the adversities that weigh on much of the African-American community appear to be inducing significant outmigration by members of this population group, and in several cases, notably Los Angeles, the number of African-Americans in the total population has actually now started to decline.

Thus, in spite of the extraordinary new dynamism and prosperity in American cities, the benefits are far from being equitably spread out. The well-qualified and the highly skilled have been exceptionally favored by recent economic trends in the United States, and those already at the pinnacle of the income frequency distribution have profited to a massively disproportionate degree. At the lower end of the labor market, by contrast, the same trends have given rise to few or no benefits whatever. Even if median incomes have been trending upward on the whole, there have been some quite significant losses, notably among the expanding underclass, the armies of the urban unemployed, and, at the very bottom of the social hierarchy, the outcasts and the homeless, Moreover, the gains and the losses have been very unequally distributed over different racial and ethnic groups. The overall diagnosis points once again, then, to a number of severe dysfunctionalities at the heart of the American city, even as the cognitive-cultural economy continues to generate unprecedented new rounds of urban wealth overall. In particular, as cities continue to expand in this manner, they also draw in more and more low-wage immigrants from the four corners of the world, thus creating over and over again prevailing patterns of inequality and exclusion.

Table 6.5. Race and ethnicity in the New York–Northern New Jersey–Long Island MSA, the Los Angeles–Long Beach–Santa Ana MSA, and the Chicago–Naperville–Joliet MSA, 2005

	Percent of total population in each MSA		
	New York–Northern New Jersey–Long Island	Los Angeles–Long Beach–Santa Ana	Chicago–Naperville–Joliet
White	58.4	53.3	64.7
African-American	18.3	7.2	18.1
Hispanic origin	22.1	43.9	19.4
Asian	9.7	14.6	5.6
Total population	17,214,000	12,723,000	8,988,000

Source: American Community Survey, accessed through IPUMS USA, Minnesota Population Center, at http://usa.ipums.org/usa/. NB: White and Hispanic-origin populations are not necessarily mutually exclusive. (MSA = metropolitan statistical area.)

Social Stratification and Spatial Segregation in the City

The patterns of light and shade that run through large American cities are nowhere more visible than in the variety of residential neighborhoods that (in terms of sheer spatial extent) constitute the greater part of the physical fabric of urban areas. There is nothing particularly new in this phenomenon of neighborhood diversity, for the social space of cities has always been characterized to a significant extent by spatial differentiation. Of late years, however, the degree of sociospatial fragmentation has evidently been increasing in large metropolitan areas by comparison with cities in the era of fordist mass production (Bobo et al. 2000). This fragmentation is in part an effect of the traditional cleavages of social class and race in American society, and in part an outcome of the intensifying inflow into American cities of immigrants who bring with them an ever-widening range of ethnic, linguistic, and cultural idiosyncrasies. These varying principles of social demarcation indicate that the American city, perhaps as never before in its history, represents today a veritable demographic kaleidoscope that is replicated, imperfectly but appreciably, in the individual neighborhoods spread out over its entire spatial extent.

Something of the character of social space in America's largest metropolitan areas can be gleaned from a glance at Table 6.5 which

121

shows the main patterns of race and ethnicity in the New York–Northern New Jersey–Long Island MSA, the Los Angeles–Long Beach-Santa Ana MSA, and the Chicago–Naperville–Joliet Chicago MSA. Nonwhites now actually account for almost half of the populations of the three metropolitan areas. The number of African-Americans in these cities remains significant (notably in New York and Chicago), but in recent years has been either stable or declining. The Hispanic-origin population is in all three cases larger than the African-American population, especially in Los Angeles, and is increasing at a rapid pace. The Asian immigrant population is relatively small, though it constitutes a definite presence in contemporary American cities, and it, too, is growing rapidly. Both the Hispanic-origin and Asian populations decompose into much more detailed subcategories based largely on national/cultural differences (Mexican, Puerto Rican, and Dominican, for example, or Chinese, Vietnamese, and South Korean) many of which also condense out in urban space into distinctive residential areas. Much of the economy of places like New York and Los Angeles would be effectively immobilized if immigration from Latin America and Asia were to be radically curtailed. Nevertheless, the fortunes of the latter two groups of immigrants often diverge greatly, in part as a consequence of their contrasting sociocultural attributes, and in part as a consequence of their differing levels of education and training. Many Asian immigrants, for example, already have an extended experience of urban life before their arrival in the United States and possess relatively high levels of formal qualification when they first enter the country. On these counts, they are on average somewhat better situated than Hispanic immigrants to gain access to jobs that provide a degree of upward social mobility.

Over the greater part of the twentieth century, immigrant neighborhoods in American cities were located principally in central areas close to the industrial facilities where their residents were for the most part employed. Today, immigrant communities are equally likely to be found in the far suburban reaches of the metropolis where numerous low-wage job opportunities now abound in burgeoning technopoles, decentralized manufacturing plants, and service industry complexes. Conversely, many inner city areas that formerly served as residential neighborhoods of the low-wage and immigrant urban labor force have been subject to an accelerating process of gentrification. Much of this process can be ascribed to the search by elite workers in the new cognitive-cultural economy for housing that provides them with ready access to jobs in central business districts and adjacent commercial zones as well as to the upscale shopping, entertainment, cultural facilities, and social scenes (themselves often precursors of further gentrification) that have proliferated in inner city areas of late years. Even so, poor neighborhoods housing both immigrant and nonimmigrant populations are still a feature of inner city areas, and these are critical sources of labor for the low-wage jobs that continue to proliferate in central business districts and in the industrial districts adjacent to them. At the same time, immigrant neighborhoods all across the city function as staging posts within complex networks of information flow that link would-be migrants in various parts of the world periphery to potential employers in the large metropolis. New immigrants are drawn in large numbers into the orbit of the city as information is exchanged backward and forward through these networks. By contrast, the African-American ghettos that form an integral element of the fabric of poor inner city neighborhoods are virtually always significantly less well integrated into the local economy—even into its low-wage segments—compared with many immigrant groups. These ghettos represent one of the most enduring and obdurate dilemmas in America today, for more than any other part of the city, they display concentrated symptoms of social breakdown, neglect, and failure that continually reinforce their relative isolation from the rest of urban society (Massey and Denton 1993; Wilson 1987).

These observable patterns of sociospatial segregation in large metropolitan areas result from a multiplicity of interpenetrating push and pull factors working through both the medium of intra-urban space and the logic of the housing market. Obviously, it is not possible in the present account to deal with the full complexity of this topic, but three special points merit attention at this stage. First of all, a variety of pressures, both positive and negative, combine to induce many households to gravitate to residential areas whose social and cultural attributes are compatible with their own. In this manner, households seek to satisfy many different needs in respect of social reproduction and the pressures of social life. In low-income neighborhoods, especially where immigrants are preponderant, these needs often flow from the imperfect assimilation of individuals into the wider society and their dependence on the community for such matters as social support or practical information about work opportunities. In high-income neighborhoods the search for prestige, peer approval, superior schooling opportunities for children, and so on, play a critical role. Second, as we know from gravity-model analyses, there is always a strong inverse correlation between population density and distance between place of residence and place of work in the city, though the relationship is also powerfully structured by income levels (Wilson 1972). In particular, low-income immigrant neighborhoods in American cities are typically located in areas lying adjacent to districts where a large proportion of their residents are able to find employment. Higher-income workers evidently enjoy a much greater range of spatial choice in regard to residential location. Third, intra-urban social segregation is reinforced by certain kinds of institutional arrangements, as well as by continuing discrimination in housing markets. Zoning laws, for example, help to create areas of social exclusion by limiting the types of housing (e.g. apartment buildings) that can be built in designated areas, or by restricting the number of families that can be housed on any single lot.

Problems of the latter type are exacerbated under conditions of municipal fragmentation, and especially by the secession of relatively wealthy communities (such as Beverly Hills in the wider context of Los Angeles) from the metropolitan polity as a whole, thereby facilitating the implementation of planning decisions that protect the special interests of local property owners. This form of secession, by the way, typically makes it difficult or impossible for the less prosperous municipalities in the wider metropolitan area to share in the tax revenues of the breakaway communities. It also limits the possibilities of imposing compensating penalties on residents of the latter communities in return for the advantages that they take of the public goods and other socioeconomic opportunities offered by the adjacent metropolitan milieu. The recent spread of gated communities in wealthier neighborhoods of American cities adds further to these discriminatory spatial practices and intensifies the undemocratic partition of intra-urban space. Urban analysts have frequently pointed out, too, that discrimination in housing markets is still widely practiced by banks, real estate agents, and homeowners, despite the provisions of the Fair Housing Act of 1968 and its subsequent amendments (Massey and Denton 1993; Young 1999). Clearly, there is always some wide range of residential choice for underprivileged social groups in American cities, and even in the best of all possible worlds, some degree of self-selected segregation would certainly occur. The point, however, is that this range of choice is also actively limited by an assortment of housing market and institutional logics, and that in a society where, as President George W. Bush is wont to repeat, freedom and democracy are upheld as sacred values, actual practice frequently falls far short of declared principles.

It is evident, then, that major questions about social justice and equity in American cities remain strongly on the agenda, even as the new cognitive-cultural economy pushes privileged social fractions and selected areas of the same cities to ever higher levels of economic well-being and physical embellishment. In the end, the pertinence of these questions can be traced back to two main lines of force that have been alluded to repeatedly in the foregoing discussion. One of these resides in the dynamics of the urban cognitive-cultural economy and the ways in which it engenders a deeply segmented pattern of material and psychic rewards from work. The other emanates from the structure of urban social space itself, and to the ways in which it helps to channel the life of urban denizens into highly differentiated zones of advantage and deprivation. These lines of force operate in mutually sustaining ways, for just as wage and salary levels shape any given individual's prospects in the housing market, so conditions of social reproduction in residential space affect the same individual's abilities to perform in the workplace. In the light of these striking social justice and equity issues, recent advocacies by promoters of the creative-city idea to the effect that best-practice policy consists in large degree of subsidizing upper-tier workers by investing in urban amenities that cater to their supposed tastes and preferences, must be viewed as being not only theoretically uninformed (see Chapter 3) but also deeply regressive on the political front. This remark is greatly reinforced when we take into account the pressing need across American cities for investment in such critical areas of social and economic development as public housing, transport services, workforce training, healthcare, and all the rest.

Citizenship and Democracy

The imperative of urban governance flows in the first instance out of the condition of the city as a spatial entity riven by peculiar tensions, collisions, and failures that threaten its overall social and economic viability, and that are irresolvable in the absence of some

125

agency or agencies of collective action and coordination. However, the institutions of urban governance are never simply just sites of technocratic management; they are also continually subject to political contestation by an assortment of social constituencies seeking to influence the outcomes of collective action in their own interests.

Urban governance, then, is focused on the provision of specific kinds of public goods and services, but always in a manner that reflects particular condensations of different political tendencies and pressures in urban space. Institutions of governance in American cities are represented by a patchwork of both formal and informal organizations including many different kinds of civil bodies. This complex skein of institutional arrangements demarcates a terrain of citizenship, not just in the trivial sense that equates citizenship with the simple fact of residence in a city, but more importantly in the sense that urban dwellers are also explicitly endowed with localized rights and bound by localized obligations (Isin 1999). These endowments establish a framework of actual and potential lines of participation in the urban community, though the spirit and purpose of citizenship are only fully consummated when the citizenry itself participates actively and energetically in bringing it to life. In theory, at least, the practice of citizenship continually mobilizes and remobilizes the democratic order of the city while at the same time helping to build the urban community as a whole. That said, prevailing conceptions of citizenship and democracy in the United States today revolve for the most part around the liberal-constitutionalist tradition à la Locke, which puts a premium on property rights, freedom of speech, competitive markets, and the franchise. This is assuredly a powerful vision, but one that also has a number of blind spots. In particular, and in view of the earlier discussion of the chronic social and economic inequalities in the city, it is a vision that is strikingly silent about social disparities, and the ways in which they distort citizenship and democratic participation in practice. In the urban arena, in particular, where the rich and the poor, the privileged and the underprivileged, the socially integrated and the totally marginalized exist cheek-by-jowl in close geographic proximity to one another, these disparities translate directly into tangible deficits and surpluses of civil prerogatives from one part of the city to another. It is scarcely a matter for surprise, given these conditions, that political engagement in American cities is less inclined to resemble a traditional New England town meeting than it is to take the form of clashes between interest groups in a sporadic war of position. One manifestation of this state of affairs is the periodic rise of popular movements in modern cities, especially as groups toward the lower end of the socioeconomic ladder seek in various ways to redress the inequalities and injustices that stand in the way of their full incorporation into the urban community.

In spite of the turbulence that is endemic to large cities, the present moment is one in which grassroots social movements seem recently to have receded from the forefront of urban political life, and in which an uneasy quiescence—interrupted by erratic popular protests—seems to hold sway. This comment is underlined by a comparison of the current conjuncture with the period of the 1960s and 1970s when massive and sustained levels of political mobilization around issues of racial and ethnic injustice, and what Castells (1976) referred to as the urban contradictions of collective consumption, were persistent elements of the social landscape of major metropolitan areas. In the context of the forces unleashed by the combined effects of the new economy, globalization, and neoliberalism, some loss of focus in urban political life is no doubt understandable, but concomitantly, the present moment is one in which significant reanalysis of the issues combined with practical regrouping and restrategizing around the urban question must be a high priority in politically progressive circles (see, e.g., Orfield 1997). In a recent book, Judis and Teixeira (2002) have offered the hopeful suggestion that a socially progressive majority is in the process of consolidating a hold over what they call "post-industrial" cities in America. This suggestion may well turn out to be substantially correct, though it calls for some qualification in relation to the main ideological points of reference of the new urban elite, or, alternatively, to what Markusen (2006) has called the political project of the creative class. Two significant markers of the sociopolitical inclination of the creative class are said (by Florida 2002, 2004, for example) to be tolerance and a taste for diversity. These dispositions are no doubt excellent in and of themselves, and a brief might possibly be set forth for the proposition that never have they been so pervasive as they are today in large American cities. Equally, a case might be made to the effect that attitudes of tolerance among the creative class in contemporary American cities often amount to not much more than indifference and detachment, while advocacy of diversity seems all too often to point toward mere mechanical juxtaposition of

127

different social groups, and, in practice, stops well short of practical engagement in free and open association with all comers. Possessive individualism and a significant measure of self-obsession are probably just as accurate as descriptors of the outlook of the new urban elite as the disinterested enthusiasm for social equity and justice that some creative-city theorists and other urban analysts ascribe to them.

In short, something very much more than the liberal nostrums of the recent past would seem to be essential if progressive urban reform is to be achieved in the current economic and social climate. What that something entails remains very much an open question at the present time, but it might perhaps in some sense be identified in terms of a revivified social democracy, or at least a political program capable of envisioning an urban future that goes beyond the notion of the creative city with its implicit celebration of a lopsided consumer society, and that is able to push toward a convivial and inclusive city for all. This implies, again, the need for a judicious balance between well-thought out local economic development programs and a far-reaching concern for the material welfare of the citizenry at large. The ultimate standard of success in the matter of urban policy, then, will not only be an economically healthy and innovative city, but also the redemocratization of urban space, the emergence of a new public sociability, and income redistribution. Significant spatial reallocation of urban public goods and services, for example, is one way in which the latter goal might be pursued. In reply to those who are no doubt ready at this point to declare that any such maneuver is destined to compromise the growth potentials of the city, we need only call attention to the circumstance that the continuing public penury and social distress that characterize so much of American metropolitan areas today are surely among the most evident obstructions in the way of a fuller flowering of the new economic order of the twenty-first century. In addition, enlargement of the sphere of urban democracy is a major social imperative, first of all because considerations of equity and social justice are of major importance in their own right, and second of all because enlargement is a significant practical means of registering and dealing with the dysfunctionalities that inevitably occur in dense social communities. This remark is based on the observation that the mobilization of voice in such communities is a critical instrument for the constructive treatment of their internal stresses and strains. Large cities, with their expanding social problems are confronted with a series of particularly urgent political challenges in this regard, for not only is their internal social stability in jeopardy, but also because any failure to act is likely to undermine the effectiveness of growth strategies posited on unleashing ever higher levels of innovation and productivity. Lastly, some further reconsideration of the notion of practical citizenship is in order. An enlarged definition of citizenship, one that is in harmony with the unfolding new world system, would presumably ascribe fundamental political entitlements and obligations to individuals on the basis of their changing involvements and allegiances in given urban and/or regional communities. As it happens, traditional conceptions of the citizen and citizenship are vigorously in question at every geographic level of the world system—for we are all of us rapidly coming to be, at one and the same time, participants in local, national, multi-nation, and global communities—but nowhere as immediately or urgently as in the large cities of the modern world (Holston 2001; Keating 2001). Even though only a few tentative and pioneering instances of pertinent reforms in such cities are as yet in evidence (as in certain countries of the European Union, for example), experiments in local political enfranchisement will no doubt come to be increasingly common in the near future as metropolitan areas start to deal seriously with the unfolding economic and political realities that they face. In a world where mobility is significantly on the rise, it is not entirely beyond the bounds of the conceivable that individuals will one day freely acquire title of citizenship in large cities many times over in conjunction with their movements from place to place throughout their lifetimes.

7

City-Regions: Economic Motors and Political Actors on the Global Stage

Introduction

Contrary to a number of recent predictions, geography is not about to disappear. Even in a globalizing world, geography does not become less important; rather, it becomes increasingly important because globalization enhances the possibilities of heightened geographic differentiation and locational specialization. By the same token, an extended archipelago or mosaic of large city-regions is currently making its historical and geographical appearance on the global stage, and these peculiar agglomerations are now beginning to function as important spatial foundations of the new world system that has been taking shape since the end of the 1970s (Scott 1998b; Veltz 1996). The internal and external relations of these city-regions and their complex growth dynamics present a number of extraordinarily perplexing challenges to researchers and policymakers alike as we enter the twenty-first century, and the challenges are underlined by the fact that many of the most important internal and external economic relations of these places are being molded by the dynamics of the cognitive-cultural economy.

There is an extensive literature on "world cities" and "global cities" by analysts such as Castells (1996), Friedmann and Wolff (1982), Hall (1966), Knox (1995), and Sassen (1991) to name just a few of the more important commentators on these topics. This literature focuses above all on a concept of the cosmopolitan metropolis as a command post for the operations of multinational corporations, as a center of information-processing activities and advanced services, and as a deeply segmented social sphere marked by extremes of poverty and wealth. I use the same concept as a basic point of departure in this chapter, but I also seek to extend its range of meaning so as to incorporate the idea of the metropolitan area along with its surrounding hinterland region as an emerging socioeconomic unit marked by ramifying local institutions and an increasingly distinctive political identity, and, concomitantly, by a growing self-assertiveness on the global stage. I designate any phenomenon of this sort by the term *global city-region* (Scott et al. 2001). In practice, global city-regions consist of enormous expanses of contiguous or semi-contiguous built-up space, often focused on a central metropolis but sometimes even taking the form of juxtaposed metropolitan areas. These core elements are in turn surrounded by hinterlands of variable extent which themselves may be sites of scattered urban settlements. The internal economic and political affairs of these metropolis–hinterland systems are bound up in intricate ways in intensifying and far-flung extra-national relationships. In parallel with these developments, embryonic consolidation of global city-regions into definite territorial-*cum*-political entities is also occurring as their component units of local government (counties, metropolitan areas, municipalities, special administrative areas, and so on) club together to form spatial coalitions in search of effective bases from which to deal with both the threats and the opportunities of globalization. So far from being dissolved away as geographic entities by processes of globalization, city-regions are actually thriving at the present time, and they are, if anything, becoming increasingly central to the conduct and coordination of modern life.

An initial and admittedly inadequate empirical identification of global city-regions in the world today can be made by reference to the map of large metropolitan areas shown in Figure 7.1. Not all large metropolitan areas, however, are equally caught up in processes of globalization, and not all global city-regions can be simply equated with existing large metropolitan areas, as we shall see with greater clarity below. Even so, the pattern revealed in Figure 7.1 plainly indicates that large-scale urbanization is not only of major importance in the contemporary world, but that it is also characteristic of economically advanced and economically developing countries alike. Moreover, large cities all over the globe continue to grow in size. In 1950, there were 83 cities in the world with populations

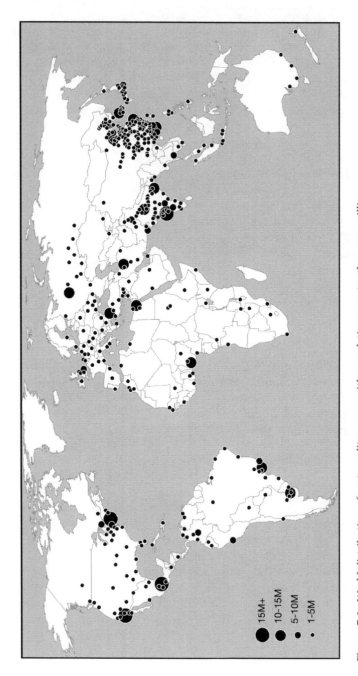

Figure 7.1. World distribution of metropolitan areas with populations greater than one million

Source to data: United Nations (2004).

of more than one million, two-thirds of them being located in the economically advanced countries. In the year 2005 there were 454 such cities, the vast majority of them now being in the economically developing countries. According to the United Nations, significant population growth can be expected to continue in the world's largest metropolitan areas over at least the next couple of decades (see Table 1.1).

Globalization and the New Regionalism

In the immediate post-War years, almost all of the major capitalist countries were marked by strong central governments and relatively tightly bordered national economies. These countries constituted a political bloc within the framework of a *Pax Americana*, itself supported by a rudimentary network of international arrangements (the Bretton Woods system, the World Bank, the IMF, GATT, and so on) through which they sought to regulate their relatively limited—but rapidly expanding—economic interrelations. Over much of the post-War period, the most prosperous of these countries could be said to constitute a core zone of the world economy, surrounded in turn by a peripheral zone of Third World nations, with a complex set of interdependencies running between the two, as described by world system theorists like Wallerstein (1979).

Today, after much technological change and economic restructuring, significant transformations of this older order of things have occurred virtually across the world, bringing in their train the outlines of a new world system or a new social grammar of space (Badie 1995). One of the outstanding features of this still emerging system is the apparent but still quite inchoate formation of a multitiered spatial structure of economic and political institutions ranging from the global to the local. Four main aspects of this state of affairs call for immediate attention:

- Huge and ever-increasing amounts of economic activity (input–output chains, migration streams, foreign direct investment by multinational corporations, monetary flows, and so on) now occur in the form of cross-border relationships. Such activity is in important ways what is generally meant by globalization as such, even though it is far indeed from any ultimate point of fulfillment. As globalization in this sense moves forward, it creates

133

numerous actual and potential predicaments that in turn set in motion a variety of political responses and institution-building efforts. Practical expressions of such efforts include international forums of collective decision-making and action such as the United Nations, the G7/G8 group, the OECD, the World Bank, the IMF, the World Trade Organization, and numerous NGOs. While these evolving political responses to the pressures of globalization remain limited in scope and severely lacking in real authority, they are liable to significant expansion and consolidation as world capitalism continues its predictable expansion.

- In part as a corollary of these same pressures, there has been a proliferation over the last few decades of multination blocs such as the EU, NAFTA, MERCOSUR, ASEAN, APEC, CARICOM, and many others. These blocs, too, can be seen as institutional efforts to capture the benefits and control the negative externalities created by the steady spilling over of national capitalisms beyond their traditional political boundaries. Because they involve only small numbers of participants, they are more manageable as political units (i.e. transactions-costs problems are relatively restrained) in comparison with actual or putative global organizations. Even so, they remain for the most part in rather elementary stages of formation at the present time, with the EU being obviously in the vanguard.

- Sovereign states and national economies remain the dominant elements of the contemporary global landscape, though they are clearly undergoing many sea changes. On the one hand, individual states no longer enjoy quite the same degree of sovereign political autonomy that they once possessed, and national economies have been subject to massive debordering over the last few decades so that it is increasingly difficult, if not impossible, to say precisely where, say, the American economy ends and the German or Japanese economies begin. On the other hand, they find themselves less and less able or willing to safeguard all the regional and sectional interests within their jurisdictions. As a result, and as noted in the first and second points above, some of the regulatory functions that were formerly carried out under the aegis of the central state have been drifting to higher levels of spatial resolution; simultaneously, other functions have been drifting downward (Swyngedouw 1997).

- Most importantly for the purposes of the present account, there has of late been a resurgence of region-based forms of economic and political organization all over the world, with the most overt expression of this tendency being manifest in the formation of large global city-regions. These city-regions make up a global mosaic that is now in many ways beginning to override the old core-periphery system that has characterized much of the macro-geography of capitalist development since its historical beginnings. In particular, global city-regions are found not only in the more advanced capitalist societies, but increasingly in many less economically advanced countries as well. In the latter case, they frequently serve as major foci of economic development impulses that then diffuse more widely across the national territory. The roots of the new cognitive-cultural production system penetrate deeply into the hearts of many of these city-regions, and it is from these staging posts, too, that much of the contemporary global system of trade in cognitive-cultural products and services is managed.

Of course, the composite political map of the contemporary world is vastly more complicated than this simple quadripartite schema suggests, for it is characterized by many interscalar and cross-scalar institutional arrangements, both governmental and nongovernmental, in addition to those described above (Hardt and Negri 2000). In addition, we need to guard against interpreting this schema in terms of a strictly ordered hierarchy, from top to bottom, of political authority and subordination; rather, the institutions identified at every spatial level have significant degrees of decision-making autonomy, and lines of influence extend upward as well as downward through the entire structure of relationships.

The fourth main point laid out above calls especially for further amplification. As we have seen in earlier sections of this book, the propensity of many types of economic activity to gather together in dense regional clusters or agglomerations appears to have been increasing greatly in recent decades. The rise of the cognitive-cultural economy over the last few decades has been an important though certainly not the only factor in this tendency. This renewed quest for collective propinquity—and its reflection in the resurgence of cities—is both an outcome of the localized increasing-returns effects that flow from the peculiar socioeconomic bases of much of the contemporary economy, and a strategic response to the pressures and opportunities of globalization. Propinquity is important because it is a source of enhanced competitive advantage for many firms, and hence regional production complexes are coming increasingly to function as territorial platforms for contesting global markets. Moreover, the diminishing commitment of national governments to resolving all the nuanced policy problems of each of the individual regions contained within their borders means that many regions are now faced with the choice of either passive subjection to external cross-border pressures, or active institution-building, policymaking, and outreach in an effort to turn globalization as far as possible to their own advantage. Regions that take the latter course are likely to find themselves also faced with many new tasks of political coordination and representation. These tasks are of special urgency at a time when large city-regions function more and more as poles of attraction for low-wage migrants from all over the world, so that their populations are almost everywhere heavily interspersed with polyglot and often disinherited social groups. As a consequence, numerous city-regions today are also being confronted with pressing issues related to political participation and the reconstruction of local political identity (see Chapter 6).

The Economic Foundations and Global Spread of City-Regions

Transactions Costs and Organizational Synergies

One of the seeming paradoxes of the field of investigation at hand is that whereas dramatic improvements in technologies of transport and communication over the last few decades have greatly reduced many spatial barriers, thus bringing all parts of the world into ever closer contact with one another, dense urban agglomerations continue to increase in size and importance everywhere. These apparently contradictory trends turn out on further scrutiny to be two faces of a mutually reinforcing set of relationships whose geographic logic can in significant ways be understood in terms of the network arrangements and relational interdependencies that constitute the basic structure of organized economic and social life in the contemporary world.

This logic reposes in the first instance on an intrinsic duality of any economic or social organization, namely, its status as (a) a system of nodes and links with definite spatial extent, and (b) a social structure marked by forms of interaction, bonding, and symbiosis. The first element of this duality signifies that any bilateral or multilateral relationship (or transaction) will tend to incur locationally dependent impedances or costs of varying intensity, depending on the nature of what is being transacted. The second element signifies that we can expect various kinds of synergistic outcomes to emerge from the collective presence of many different agents within the network. Examples of these synergies might be the cost- and risk-reducing effects of just-in-time linkage arrangements in input–output systems, or the information that accumulates and circulates in local labor markets about job opportunities for workers, or the knowledge spillovers that occur in day-to-day business dealings between individual firms. On many occasions, these synergies themselves are sensitive to the effects of geographic space, in the sense that their intensity increases or decreases depending on the distance that lies between the parties involved. Their force is often at a high level when many interdependent social and economic activities are located together in a relatively confined geographic area, in which case, they take on the explicit form of *agglomeration economies*, or localized increasing-returns effects (see the discussion in Chapters 1, 3, and 4). Note, in addition, that the spatial clustering of social and economic activities will always tend to be intensified by the presence of large-scale infrastructural artifacts, which are, in effect, a further kind of agglomeration economy.

Before we proceed on, a further word of clarification about the distinction between transactional relationships and organizational synergies is required. As implied in the previous paragraph, these two phenomena are not wholly independent of one another, for the former often constitute channels through which the latter are diffused and appropriated. Consider a case not unlike the classic example of the relations between bee-keeping and apple production that Meade (1952) has proposed. Suppose that factory *a* procures inputs from factory *b*. This is a transactional relationship, and one that is, of course, subject to distance-dependent costs. We might possibly say that on account of this relationship, factories *a* and *b* represent simple externalities for one another. However in this analysis, I want to exclude traded interdependencies from any notion of externality and to account for them only on the input–output side of the ledger. Suppose now that this input–output relationship is also associated with useful—and gratis—information flows from *a* to *b* and/or from *b* to *a*. The information received in this manner represents a pure externality or synergy. In view of these definitions, we can say that there is a twofold incentive in terms of spatial transactions costs and externalities for factories *a* and *b* to locate close by one another. On the one hand, a diminution of the distance between the two will result in lower transactions costs to the benefit of at least one of the relevant parties. On the other hand, the same diminution is likely to be associated with increased intensity and frequency of useful information flows. Of course, externalities can sometimes carry a negative charge, in which case there would be an incentive for the relevant parties to locate at some distance from one another, but this complication need not detain us for the present. The location of production activities can thus be approached, in a first and rather abstract formulation, in relation to the combined and interdependent effects of transactions costs and externalities. At a second and more concrete level of analysis, we also need to acknowledge that these effects unfold in empirical terms in enormously complex and varied ways depending on the specificities of technology, market structure, prevailing patterns of industrial organization, and so on.

The Emergence of Urban Superclusters

This spatial and organizational logic plays a fundamental role in shaping the economic landscape of capitalism, and in particular, in regulating the degrees of locational agglomeration and dispersal that occur in any given historical and geographical instance. Obviously, if transactions costs are consistently low and latent agglomeration economies are largely absent, we would expect to observe relatively high levels of locational dispersal, even among functionally interrelated producers; conversely, if transactions costs are high and agglomeration economies are well developed we would expect relatively intense levels of locational clustering to occur. In fact, we can construct a number of very much more complex scenarios than these by observing how locational structures vary in response to a series of graduated shifts in the intensity of transactions costs and externalities. A systematic attempt to codify the ways in which these two variables interact with one another to generate contrasting locational patterns and economic landscapes is laid out in the Appendix to this chapter. Our immediate concern for the present is to pinpoint some of the main ways in which these variables seem to be operating today, and above all within the context of the cognitive-cultural economy and its various appendages.

Two important observations now need to be advanced. First, transactions costs in contemporary capitalism tend to be extremely variable as a function of distance, depending on both the substantive content of the transactions themselves and the nature of the channels through which they flow. This state of affairs is dramatically illustrated by the internal and external linkage structures of production centers like Silicon Valley, Hollywood, or the City of London. Within any one of these clusters, firms are often tightly linked together in critical networks of high-cost transactional relations (many of them involving face-to-face contacts), whereas much of the output of the same firms circulates with relative ease around the globe (Scott 2005). Second, opportunities for appropriating localized increasing-returns effects appear to have intensified greatly in many segments of the economy since the advent of the cognitive-cultural turn in the waning decades of the twentieth century. We have already noted, for example, that the emergence of flexible, post-fordist production and employment systems has been attended by a proliferation of social interactions within interindustrial networks and local labor markets. These interactions then underpin an enormous diversity of information spillovers, learning effects, and innovative impulses (Cooke and Morgan 1998; Maskell and Malmberg 1999; Scott 2006a). Accordingly, wherever we find the peculiar combination of circumstances described above—high transactions costs at the core of the production system, low transactions costs on at least an important share of final products, and strongly developed agglomeration economies—there we are also likely to detect propitious conditions for the materialization of economic and social superclusters (see Appendix). Much the same idea can be expressed by saying that the intersecting effects of high transactions costs and localized increasing returns at the functional hubs of flexible (and notably cognitive-cultural) production complexes *together with* transactional conditions that make it possible to serve large-scale markets for final products, provide a number of essential economic foundations for the emergence of modern global city-regions. The operation of these effects on local growth is magnified when not just one but several different individual complexes of productive activity occur within the boundaries of a single metropolitan area.

In the concrete circumstances of contemporary capitalism, then, cities are not only resurgent, but are also taking on hitherto unprecedented dynamics and physical form. Recall that some of the most important building blocks of the modern economy are represented by cognitive-cultural sectors like technology-intensive production, business and financial services, neo-artisanal and fashion-oriented manufacturing, cultural industries, the media, and so on. Sectors like these are eminently susceptible to the local–global spatial logic outlined above, and, to be sure, they frequently function as instruments of accelerated growth in contemporary global city-regions as well as in more specialized kinds of clusters. Growth of this sort, moreover, is often inscribed in a peculiar dynamic of self-reinforcement, for when markets widen and clusters expand, agglomeration economies tend to increase in response to deepening organizational complexity at the point of production, thus creating further possibilities for market contestation, in successive rounds of interdependence. Even if urban growth is sometimes accompanied by countervailing trends to decentralization (as in the case of runaway production from Hollywood or the offshore movement of semiconductor assembly plants from Silicon Valley), there is little concrete evidence hitherto to suggest that the vigorous growth characteristics of superclusters, as identified, are generally on the point of collapse. To the contrary, and especially in instances where the new cognitive-cultural economy is deeply ensconced, global city-regions are virtually everywhere continuing to grow in terms of production and population, and to expand outward in terms of their spatial extent.

The Global Mosaic of City-Regions

Large cities or city-regions, then, have today become a notably insistent element of the geographic landscape. Throughout the world, numerous suitably positioned metropolitan centers together with their surrounding hinterlands are evolving into superclusters whose development stems from the circumstance that so many of the leading sectors of capitalism at the present time are organized as intensely localized networks of producers with powerful endogenous growth mechanisms and with increasingly global market reach. In cases where individual metropolitan areas lie in close proximity to one another, they may actually fuse together to form yet bigger global city-regions, as exemplified by such huge agglomerations as the greater New York, Los Angeles, or Tokyo urbanized areas. Each city-region is the site of intricate webs of social and economic inter-dependencies, and each is the locus of robust endogenous growth dynamics powered by expanding external markets. As such, city-regions are not only of great size, but also growing constantly larger (cf. Figure 7.1). In many respects, these entities can be thought of as regional motors of the new global economy, that is, the prin-cipal sites of production, economic growth, and innovation in the world today. As such, they are also typically caught up in intense interrelationships of mutual trade and exchange entailing not only high levels of intercity competition but, in addition, many different kinds of collaborative undertakings, joint venturing activity, and financial interdependencies. They constitute an expanding mosaic or archipelago that now extends across the whole of the world (Scott 1998b; Veltz 1996).

All of that being said, wide expanses of the modern globe—former colonies, ex-socialist states, areas occupied by traditional cultures unreceptive to capitalist norms of economic and social life, and so on—remain at the extensive economic margins of capitalism and are often stubbornly resistant to development. Even so, these parts of the globe are sometimes punctuated by islands of relative prosperity and opportunity in the guise of burgeoning urban centers, and some of these are undoubtedly aligned along an evolutionary pathway that will lead them eventually to much higher levels of economic growth. In the 1960s and 1970s, places like Hong Kong, Singapore, Seoul, and Mexico City were positioned at various points along the early stages of this pathway, but all of them have now moved decisively into positions of much more advanced development. Today, many urban regions (e.g. Bangkok, Guangzhou, São Paulo) in a diversity of low- and middle-income countries are following on the heels of these pioneers, while parts of, say, Nigeria, Indonesia, and Vietnam seem poised to follow suit. According to an older dependency theory school of thought as expressed by analysts like Amin (1973) and Frank (1978), this sort of advanced development was never sup-posed to happen in the world periphery. More astonishing still is the emergence of significant cognitive-cultural production systems in major cities of erstwhile underdeveloped countries, even as many of these cities continue to function as sources of cheap labor within global commodity chains controlled by manufacturing firms in the United States, Europe, and Japan. The cases of Hong Kong, Shanghai, Bangkok, Singapore, Seoul, Beijing, Bangalore, Mexico City, and Buenos Aires, among others, are noteworthy here. A number of these cities are working strenuously at the present time to establish policy frameworks that will enhance their role as centers of the new creative economy and that will reinforce their presence on the global stage.

The Political Organization of Global City-Regions

The space-economy of the modern world is thus currently in a state of rapid flux, leading in turn to many significant adjustments in patterns of political geography. On the one hand, the profound changes that have been occurring on the economic front are giving rise increasingly to diverse responses and experiments in regulatory coordination at different geographic levels from the global to the local. On the other hand, the new regulatory institutions that are beginning to assume a clearer outline on the world map help to channel economic development into spatial structures that in part run parallel to the ordered tiers of political–geographic outcomes as described earlier. The political shifts going on at each of these tiers pose many perplexing problems, but the level that is represented by the new global mosaic of city-regions is particularly enigmatic and is certainly one of the least well understood.

Precisely because the individual regional units at this latter level constitute the basic motors of a rapidly globalizing production system, much is at stake as they steadily consolidate their collective identities and economic foundations. The observable institutional and political ferment currently proceeding in many of these regions is intimately bound up with a search for structures of governance capable of dealing with the intensifying social and economic dilemmas that they face as their internal complexity and size increase and as their external relations become ever more extended. A major component of this search consists in efforts to build structures of coordination in the interests of orderly spatial development and to secure local competitive advantages in relation to the wider global setting. At several different stages in this book, I have offered the argument that while agglomerated socioeconomic systems are invariably axes of multiple synergies or externalities, these will always exist in some markedly suboptimal configuration if markets are not complemented by appropriate coordinating institutions and mechanisms for the supply of public goods. In putative city-regions, the imperative of overall coordination is obviously pressing in regard to the provision of large-scale, capital-intensive infrastructural artifacts such as intra-urban transportation networks, sewer systems, and water supply, as well as crucial municipal services like police, fire protection, and environmental regulation. As we have learned, however, significant structural failures also occur right at the functional core of the urban economy. These failures offer many challenges to policymakers at the best of times, but institutionally fragmented city-regions are especially ill-equipped to mount suitable responses to these challenges. Effective managerial coordination of this field of remedial opportunities is therefore of major importance, above all in a world where the continued spatial extension of markets brings each city-region into a position of vastly expanded economic possibilities, but also of greatly heightened economic threats from outside. We can accordingly expect to see much further effort invested in institution-building directed to the promotion and coordination of the economies of major city-regions as the latter continue to grow and as the predicaments brought on by globalization continue to multiply apace.

We may well ask, as a corollary, how these regions might be defined as territorial entities with greater or lesser powers of coordinated action. In many instances, no doubt, the spatial cores of city-regions can be simply and satisfactorily equated with already existing metropolitan areas. But how far out into its hinterland will the political mandate of any city-region tend to range? And how is the institutional geography of these regions to be identified when several different metropolitan areas begin to coalesce with one another, as, for instance, in the case of the northeast seaboard of the United States? These questions are in fact moot as abstract matters of *a priori* principle, though we can—drawing on a traditional marxian approach to the definition of social class—provide some methodological guidelines about how they might actually be answered in any given instance. These guidelines may be summarized by allusion to the twin notions of *objective conditions* and *political practices*. The first of these notions refers in the present context to the necessary foundation of any given city-region in a large, dense, polarized (or multipolarized) agglomeration of capital and labor with definite internal synergies and at least some degree of integration into the world system. The second refers to the active construction of territorial coalitions—whether imposed from above or coming into being spontaneously from below—in which different geographic entities (say, local government units) join together in the quest for a heightened regional capacity to deal with the administrative and policy problems that continually bubble up within these peculiar vortexes of sociospatial relationships. There is no necessary single form that these coalitions can or will take. Much depends on the circumstances of local history and geography, as well as upon issues such as efficient institutional size (as affected by intra-organizational transactions costs, for example), and every possible outcome from an overall unified structure of local government to loose intermunicipal cooperative arrangements is conceivable as a *modus operandi*. In the light of these comments, the final political–geographic outlines of any given global city-region must remain largely indeterminate in advance of concrete political mobilization. But we can already perhaps see aspects of the shape of things to come in some of the new streamlined regional government systems that have been put into place in various parts of the world over the last couple of decades, and in reference to which, Portland and Toronto are frequently cited as examples (Abbott 1997; Courchene 2001). Other intimations of possible future developments appear in the experiments and maneuvering (some of which may bear fruit, some of which will certainly lead nowhere) that are currently gathering steam around prospective large-scale municipal alliances (many of them involving trans-border agreements) such as San Diego–Tijuana, the Trans-Manche Region, the Øresund Region, Padania, Singapore–Johore–Batam, or Hong Kong–Shenzen (cf. Giordano 2000; Hospers 2006; Keating 2001). One of the most impressive of these initiatives is the current move to amalgamate Shanghai with the adjacent provinces of Jiangsu and Zhejiang to form the Yangtze River Delta metropolitan area, a gigantic global city-region of over 90 million people.

With the expansion of the global mosaic of city-regions and the propensity of each unit of the mosaic to acquire a distinctive political identity and a sense of its own collective agenda, a further series of questions starts to arise. At the very least, rising levels of concerted regional activism can be expected to lead to specific kinds of politicization and destabilization of interregional relations, both within and across national boundaries. One way in which such predicaments already manifest themselves is in the formation of regional alliances (such as the Four Motors of Europe Program established in 1988, or the proposed but now abandoned merger of the London and Frankfurt stock exchanges), leading to fears about unfair competition on the part of those excluded. Another resides in the currently prevalent attempts by representatives of some regions to lure selected assets of other regions into their own geographic orbit, often at heavy overall social cost. Yet another can be deciphered in the development races that occur from time to time when several different regions push simultaneously to secure a decisive lead as the dominant center of some budding industry, leading potentially to significant misallocation of resources. In view of the likelihood that stresses and strains of these types will be magnified as the new regionalism takes deeper hold, the need for remedial action at the national, plurinational, and even eventually the global levels of political coordination becomes increasingly pressing in order to establish a framework of ground rules for the conduct of interregional competition and collaboration (including aid to failing cities and regions) and to provide appropriate forums for cross-regional problem-solving. The European Committee of the Regions, established under the terms of the Maastricht Treaty, can be seen as an early even if still quite fragile expression in the transnational sphere of this dawning imperative.

Beyond Neoliberalism

The complex trends and tendencies alluded to in these pages pose once again the major question as to what macro-political or ideological formations will be liable to assert a dominant role in calibrating the frameworks for institution-building and policymaking at various spatial scales in the new world order. Giddens (1998) has forcefully argued that two main contending sets of political principles are now moving toward a decisive moment of collision with one another in relation to events on the world stage, certainly in the more economically advanced parts of the globe. One of these is a currently dominant neoliberal view—a view that prescribes minimum government interference in and maximum market organization of

145

economic activity (and that is sometimes but erroneously taken to be a virtually inescapable counterpart of globalization). In the light of what I have written above about the urge to collective action in contemporary capitalism, neoliberalism, certainly in the version that crudely advocates *laissez-faire* as a universal panacea, strikes me as offering a seriously deficient vision, in both technical and political terms. The other is a renascent social democratic approach. On the economic front, social democracy is prepared to acknowledge and to work with the efficiency-seeking properties of markets where these are consistent with standards of social fairness and long-term economic well-being, but to intervene selectively where they are not. As such, a pragmatic social democratic politics would seem to be well armed to face up to the tasks of building the social infrastructures and enabling conditions (at every geographic level) that are each day becoming more critical to high levels of economic performance and social vitality as the new world system comes increasingly into focus. At the city-region level, these tasks can be centrally identified with the compelling social imperative of promoting the local levels of efficiency, productivity, and competitiveness that markets alone can never fully secure, while ensuring that issues of social justice and equity are at the forefront of the policymaking process.

Globalization, then, has both a regressive side and a more hopeful, progressive side. Insistent globalization under the aegis of a triumphant neoliberalism would no doubt constitute something close to a worst-case scenario, leading over the long run to increased social inequalities and tensions within individual city-regions, and exacerbating the discrepancies in growth rates and developmental potentials between successful and unsuccessful regions all across the world. If the analysis presented here turns out to be in principle broadly correct, however, an alternative and attainable form of globalization can be envisaged, one that serves equally the goals of rising economic prosperity and progressive social reform. The emergence of city-regions in an expanding global economy offers an important occasion for rethinking many of these issues and for highlighting some of the contingent benefits of globalization. At this stage in history, the onward course of the complex structure of urbanization and globalization as described here is still quite open-ended, and it will certainly be subject in the future—both in whole and in part—to different lines of political contestation, some of which will mold it in decisive ways. In particular, and as I have tried to indicate, the dynamics of this structure raise important new questions about economic governance or regulation at all spatial levels, and some form of social market politics would seem to offer a viable, fair, and persuasive way of facing up to these questions. In the era of the cognitive-cultural economy, all the evidence points to the likelihood that city-regions will find themselves in the vanguard of the search for this brave new world.

Appendix: Locational Outcomes as a Function of Spatial Transactions Costs and Agglomeration Economies

The following diagram shows six main sets of locational outcomes, identified in relation to the horizontal and vertical axes. In the horizontal dimension, the spatial costs of transacting (per unit of transactional activity) are graded into three main categories, that is, uniformly low, mixed low and high, and uniformly high. The vertical axis refers to states of the world where agglomeration economies or localized increasing returns effects are either high or low.

<table>
<tr><td colspan="2" rowspan="2"></td><td colspan="3">**Spatial costs of transacting**</td></tr>
<tr><td>Low</td><td>Mixed low and high</td><td>High</td></tr>
<tr><td rowspan="2">**Agglomeration economies**</td><td>Low</td><td>*A*

Spatial entropy</td><td>*B*

Intermediate states between *A* and *C*</td><td>*C*

Weberian-Löschian landscapes</td></tr>
<tr><td>High</td><td>*D*

Small interconnected clusters</td><td>*E*

Superclusters</td><td>*F*

Small disconnected clusters</td></tr>
</table>

Six brief comments on the contents of the diagram are now made:

Panel A: The extreme case of spatial entropy (randomness) of all locational activities occurs when spatial transactions costs are zero and when agglomeration economies are zero.

147

Panel B: With mixed high and low transactions costs combined with low agglomeration economies, the economic landscape assumes various intermediate states between A and C, that is, a mixture of randomness and Weberian–Löschian order, where the latter term refers to simple transport-cost minimizing principles of location.

Panel C: With high spatial transactions costs but no agglomeration economies, the space-economy will be describable in terms of Weberian–Löschian landscapes, that is, spatial systems in which economic activities locate at points that minimize the distance-dependent costs of procurement and distribution.

Panel D: If spatial transactions costs are on the whole low while agglomeration economies are high, small but interconnected agglomerations will tend to occur. Producers will agglomerate because of the joint availability of agglomeration economies though only in relatively small numbers because the consistently low linkage costs will facilitate transacting activity (and many of the spillover effects that are carried along with transactional exchanges) over long distances.

Panel E: The most important case for our purposes here is represented by the situation where the transactional system is composed of a mix of high-cost and low-cost interdependencies, and where agglomeration economies are persistently high. The net outcome in this instance will tend to be the emergence of superclusters. This result will be especially well developed where there is (a) a proliferation of high-cost transactions (such as face-to-face linkages) reinforcing the clustering together of many interrelated producers, in association with (b) low-cost transactions on final products enabling producers to command distant (and in the limit global) markets.

Panel F: Where spatial transactions costs are on the whole high and agglomeration economies are also high, small disconnected clusters will tend be to observable (as in the case, for example, of proto-industrial craft communities). The presence of agglomeration economies will encourage the formation of clusters but the generally high transactions costs will make it difficult for any given cluster to grow because of limited access to external markets and resources.

Needless to say, these comments abstract away from a great many other pertinent details, including the prior spatial distribution of basic resources and fixed capital assets, but in their broad form they represent a basic point of departure for any more detailed consideration of the geometry of the economic landscape of capitalism.

8

Coda

Many different themes converge together in this book. The story that I have tried to recount is one that explores the interdependencies between the economy and urbanization in general, and that then goes on to show how these interdependencies assume very specific expression in the light of two major developments in the contemporary world, namely, the rise of cognitive-cultural capitalism and the intensification of globalization. In pursuit of these aims, I have at the outset delineated a basic concept of the urban as an explicitly spatial assemblage of human activities whose primary moment of genesis can be traced to the proclivity in capitalism for selected units of capital and labor to aggregate together in geographic space. To be sure, this moment of genesis is not the same thing as the totality of the city, and a wide variety of other social phenomena— including the political and the cultural—are also critical to the full-blown emergence of the city as a viable organism. This is in part why I speak throughout of a *social* economy of the metropolis. The notion of the social economy is all the more compelling at the present time when so much of the production system is being reconstructed around the cognitive and cultural assets of the labor force, and when so many of the sectors that compose the new economy are concentrated in large metropolitan areas around the world. In this context, too, the interconnection between the processes of production and social reproduction in the urban arena become ever more obvious. A further level of complexity is added by the deepening of globalization processes, and, as metropolitan areas become more and more entwined in these processes, the emergence of a worldwide mosaic of city-regions that function as the basic dynamos and nerve centers of the world economy. In this developmental scenario, cities are as critical to the overall success of the economy as the economy is to the birth and survival of cities.

It bears repeating here that among the few enduring generalizations we can make about the nature of cities—and perhaps the most pregnant, in terms of its power to evoke a specifically urban-centric problematic—is the notion that the urban is essentially constituted by dense systems of locational activity and land uses together with concomitant processes of social interaction organized around a common center (and associated subcenters) of gravity. It is important not to confuse the emergent effects of this phenomenon with its essence. No matter how concentrated or spread out any given city may be, its basic ossature can always be described in terms of some variation on this basic theme. As laconic as this claim may be, it is in practice highly charged with theoretical and descriptive significance. It points, in particular, to the intense interdependencies that exist among all the different entities that come together in the urban sphere and that are associated with diverse demands for mutual proximity and accessibility (as well as subsidiary demands for selected kinds of spatial separation). These demands lead, in turn, to the differential social and economic valuation of urban space as a function of location, and to the sorting out of land uses into identifiable precincts or quarters, as represented, most notably, by intra-urban industrial clusters marked by peculiar combinations of productive activity and residential neighborhoods dominated by given kinds of socioeconomic groups. All urban questions and problems, as such, bear some necessary relationship, however mediated, to these basic conditions. At the same time, the specific qualitative attributes of individual cities vary widely from one case to another, because this bare fabric of relationships is always infused with patterns of work and life whose features are, in the end, a result of the intersection of unique local circumstances with the wider social and economic environment.

Today, these attributes are being increasingly shaped by the new cognitive-cultural economy together with its associated effects on the social life and the physical form of cities. Some of the most dramatic instances of urban growth and development today can be ascribed to the peculiar features of this new economy with its extraordinary capacity for transforming technical knowledge, information, symbolic references, and cultural resonances into a proliferation of sellable goods and services. The burgeoning cognitive-cultural economy originates above all in the revalorization of workers' ingenuity in production in combination with skills-enhancing digital technologies. It thus demands from workers very different styles of engagement in the production process by comparison with the traditional smithian division of labor, as well as very different styles of labor management by comparison with the blunt relations of authority and subordination that have characterized many earlier kinds of employment regimes. Even in the bottom tier of the labor force, workers are nowadays being called upon more and more to mobilize their mental and social endowments in addition to their purely physical capacities. We might well argue in view of these trends, and above all in view of the ways in which technology, production, culture, and the symbolic realm are coming increasingly into alignment with one another in the modern economy, that the historical realization of the human community of interpretation posited by Royce (1913) is now for better or worse in the thrall of capitalist developmental dynamics. At the same time, we need to remain cognizant of the circumstance that this state of affairs by no means signals the demise of open-ended social imaginaries; if anything, it points to possibilities for working out a multitude of new ways of viewing the world and of politically negotiating out the future of the version of capitalism that is now unfolding around us.

As the cognitive-cultural economy has grown and developed over the last few decades, it has come to rest above all in major metropolitan areas of the world system. To be sure, these areas usually still harbor many routinized and standardized types of productive activity, though as shown in Chapter 3, these types of activity nowadays tend to play a diminishing role in the very largest cities, at least in so far as the United States is concerned. Equally, as we move from the top to the bottom of the urban hierarchy, as well as out into more peripheral areas, more routinized and standardized varieties of production become—with occasional exceptions—increasingly dominant. There are nonetheless many small urban centers all over the world where the cognitive-cultural economy is currently well ensconced, notably in cases where there are local traditions, or crafts, or some repository of cultural assets that can be commercialized for wider markets. Should current trends continue, it seems more than likely that yet greater incursions of cognitive-cultural producers will occur at locations further and further down the urban hierarchy.

151

The same trends, *mutatis mutandis*, are also helping to usher in the massive city-regions of the global epoch. These latest avatars of large-scale metropolitan development represent a resurgent urbanization whose fortunes contrast markedly with the experience of large cities in the waning regime of fordism in the 1970s. Cities in those years were subject to pervasive processes of capital flight and job loss that left many erstwhile prosperous regions in a state of near-devastation. By contrast, some of the most dynamic sectors in capitalism today are built upon technological, organizational, and social foundations that encourage renewed locational concentration and urban growth. In addition, the large metropolitan areas that are the most visible manifestation of this state of affairs provide ready platforms from which producers can scan, contest, and export to global markets. Constantly improving technologies of transport and communication make it possible for this structure of intertwined local and global relations to operate at high levels of efficiency while simultaneously helping to reinforce localized competitive advantages. The reduction of political barriers to trade that has gone on over the last few decades also underpins these outcomes. The overall effect is a notable resurgence of cities, both in the developed world, and to a notable degree, in the less developed world as well.

As we enter the twenty-first century, then, a new global economic and urban geography is clearly taking shape. The contemporary rise of the great bellwether centers of the global system is posited above all on the expansion of the new economy with its diversity of dynamic sectors such as high-technology electronics, biotechnology, medical instruments, software, financial and business services, banking, film and television-program production, music recording, multimedia, fashion clothing, and the like—not to mention the role of these centers as sites for conventions, sporting events, festivals, museums, galleries, concert halls, and so on. These bellwether cities are not only major hubs of economic activity in their own right, but also important poles of artistic production and experimentation with an influence that radiates across the world. Their economic and cultural weight is further enhanced by the distinctive communal sensibilities that oftentimes come into existence in response to the symbiotic social interactions that typify the round of daily work and life in given places. This distinctiveness helps to infuse unique substantive and stylistic attributes into the products of the local economy, thereby boosting the competitive advantages of each individual place. Furthermore, the general economic development model based on agglomeration economies (hence localized increasing-returns effects), monopolistic competition (hence a measure of resistance to economic contestation from other centers), and expanding/diversifying world markets (hence propitious conditions for continued local growth) leads to the plausible inference that globalization is not heading in the direction of pervasive social and spatial uniformity but will tend increasingly to be expressed in rather variegated economic and cultural patterns of urbanization.

Not all cities in this developmental regime will be unfailingly successful. However, effective governance and policymaking structures—including some capacity to achieve decision-making coordination across the urban community as a whole—will enable many cities to improve their economic prospects and possibly, on occasions, to stave off looming crises. In fact, cities of all sizes and types are confronted with numerous challenges, both old and new, as they seek to deal with their internal problems and their intensifying external relations. The endemic collisions and breakdowns within the urban land nexus call endlessly for collective supervision, even if the substantive content and intensity of these problems vary greatly from time to time and place to place. In addition to the enduring need for managerial control of problems like congestion, pollution, land use conflicts, neighborhood decay, suburban sprawl, and so on, two notably stubborn policy dilemmas have come strikingly to the fore in large cities everywhere in relation to the new economy and its increasingly global reach. One is the heightened need for appropriate modalities of collective action capable of securing localized competitive advantages and of ensuring that global threats to local economic well-being are reduced to the greatest extent possible. The other springs from the segmented socioeconomic patterns of large cities, and above all from the vast divergences in incomes and life chances that exist between the residents of affluent and poor neighborhoods. These divergences are exacerbated in a world of cities marked by ever more varied ethnic and racial composition; they are a constant source of friction and a focus of political agitation, both spontaneous and organized; and they are the basis of occasional and sometimes devastating mass outbreaks of violence.

The social divisions that characterize large cities all around the world at the present time have tended further to intensify in recent years as neoliberal ideology and practice have taken deeper and deeper hold in both the economic and political spheres. Despite neoliberal claims as to the universal efficacy and benevolence of markets, the negative consequences of the current dispensation are notably discernible in the guise of the extraordinary contrasts between the opulence and squalor that are so evident in many large American cities today. To be more specific, even when markets are working normally, cities are places where massive inequalities, irrationalities, social conflicts, and inefficient forms of lock-in appear incessantly on the horizon. As a consequence, three main types of urban policy and planning initiatives take on special urgency at the present time. These revolve around the social drive for coordinating agencies to harvest localized competitive advantages in the new urban economy, the need to build mechanisms for mitigating the democratic deficit in large urban communities, and the strategic imperative of overcoming the mismatch between the structure of intra-urban space and the institutions of urban governance. I should add that if the analysis presented here has any meaning at the end of the day, the construction of policy agendas in the search for the way forward can never be reduced simply to a matter of abstract norms or procedures, much less to chiliastic visions of ideal states of the world, but must proceed in the context of a clear feel for the possibilities and limitations of collective action in relation to prevailing social realities and frameworks of political mobilization.

Cities all around the world are bound up, too, in deepening webs of interaction with one another that pose many difficult policy dilemmas at the interurban scale. In spite of this trend, only limited forms of institutionalized harmonization among cities at the international level are currently in evidence, and the European Committee of the Regions appears to be one of the few extant institutions with a mandate specifically to coordinate inter-urban relations across different countries. The notable worldwide expansion of official and semi-official intercity networking that has been occurring of late years only partially fills this vacuum. This expansion is manifest in the multiplication of relationships such as inter-municipal contacts, exchanges of business delegations, trade-promotion exercises, cultural missions, and so on, and these sorts of relationships will almost certainly intensify in the future. In the same manner, ever-rising levels of business joint partnering and other forms of private cooperation across the cities of the world system can be anticipated.

In view of the ferment in the collective and political constitution of large urban areas, there is a certain sense in which we might say that cities nowadays represent exceptional public laboratories—both potentially and actually—for the formulation of strategies of public management and the design of civil order. Certainly they appear to have taken on a role in this regard that was relatively subdued at earlier moments in the history of capitalism. Some of the more significant instances of this type of experimentation focus on attempts to promote local economic growth, to sustain the social and political life of the community (not only by means of governmental activity, but also via city-based NGOs and other institutions of urban society), and to raise overall standards of environmental quality and urban design. All such efforts have taken on special significance with the expansion and intensification of the global cognitive-cultural economy, though a great deal remains to be achieved in terms of practical results. First, local economic development policies are still for the most part not much more than hopeful formulas that more often than not lack significant and long-term administrative backup. Second, the segmentation of urban society continues to present festering problems in both the more and the less developed parts of the globe and the democratic deficit in large urban areas is a persistent cause for deep concern. Third, large segments of the physical and environmental fabric of many cities, even in some of the world's richest countries, are locked into a state of advanced decay. Despite its numerous attractions and consolations, the city remains a deeply problematical element of modern society.

These remarks bring us back full circle to the urban question in the twenty-first century, and to the imperative of a renewal of urban theory and political practice. This imperative is conspicuously urgent in the light of the vast sea changes currently under way in contemporary cities. Some four decades ago, Henri Lefebvre (1970) made reference to the rising importance of "urban society," signifying a state of affairs where human existence is fully and finally incorporated into the sphere of the city (though in view of what I have written earlier, the notion of an urban society can never capture all the nuances of society as a whole). In today's world, economic prosperity and the human condition are even more intimately bound up with the course of urbanization than at the time when Lefebvre offered his diagnosis. We are now at a stage in human development when urban society appears to have established its ascendancy not just in a few economically advanced countries, but virtually across the globe. This remark points in practice to the ultimate erasure of any meaningful sociological distinction between the urban and the rural, or, as the case may be, the town and the countryside, or city-dwellers and peasantry. We have already, in the era of cognitive-cultural capitalism, entered into a historical and geographical moment such that the axes of the world system are in major measure defined by large city-regions that function more and more as the concentrated pivots of economic and social order and as the central reference points of symbolic value. In view of the extraordinarily powerful logic and dynamics currently at work, it seems safe to say that cities will continue to grow and that globalization will continue its onward sweep, at least for the foreseeable future. As a corollary, we can expect to see further consolidation of the global mosaic of city-regions as a basic platform of social and institutional development. The ultimate driving force behind all of these developments is the capitalist economy with its compelling urge to accumulation and self-expansion.

References

Abbott, C. (1997). "The Portland Region: Where City and Suburbs Talk to Each Other—and Often Agree", *Housing Policy Debate*, 8: 11–51.

Adorno, T. W. (1991). *The Culture Industry: Selected Essays on Mass Culture.* London: Routledge.

Amin, A., and Thrift, N. (2002). *Cities: Reimagining the Urban.* Cambridge: Polity.

Amin, S. (1973). *Le développement inégal: essai sur les formations sociales du capitalisme périphérique.* Paris: Les Éditions de Minuit.

Angel, D. P. (1991). "High-Technology Agglomeration and the Labor Market: The Case of Silicon Valley", *Environment and Planning A*, 23: 1501–1516.

Arai, Y., Nakamura, H., and Sato, H. (2004). "Multimedia and Internet Business Clusters in Central Tokyo", *Urban Geography*, 25: 483–500.

Autor, D. H., Katz, L. F., and Kearney, M. S. (2006). "The Polarization of the US Labor Market", *American Economic Review*, 96: 189–194.

——Levy, F., and Murnane, R. J. (2003). "The Skill Content of Recent Technological Change: An Empirical Exploration", *Quarterly Journal of Economics*, 118: 1279–1333.

Badie, B. (1995). *La Fin des territoires.* Paris: Fayard.

Bagnasco, A. (1977). *Tre Italie: la Problematica Territoriale dello Sviluppo Italiano.* Bolgna: Il Mulino.

Banham, R. (1960). *Theory and Design in the First Machine Age.* London: The Architectural Press.

Bardach, E. (1996). *The Eight-Step Path of Policy Analysis.* Berkeley, CA: Berkeley Academic Press.

Bassett, K. (1993). "Urban Cultural Strategies and Urban Regeneration: A Case Study and Critique", *Geoforum*, 25: 1773–1788.

——Griffiths, R., and Smith, I. (2002). "Cultural Industries, Cultural Clusters and the City: The Example of Natural History Film-Making in Bristol", *Geoforum*, 33: 165–177.

Batt, R., Christopherson, S., Rightor, N., and Jaarsveld, D. V. (2001). *Net Working: Work Patterns and Workforce Policies for the New Media Industry.* Washington, DC: Economic Policy Institute.

Baudrillard, J. (1968). *Le système des objets.* Paris: Denoël/Gonthier.

References

Becattini, G. (1987). *Mercato e forze locali: il distretto industriale.* Bologna: Il Mulino.

Beck, U. (1992). *Risk Society: Towards a New Modernity.* Newbury Park, CA: Sage.

——(2000). *The Brave New World of Work.* Cambridge: Polity Press.

Bell, C. (1924). *Art.* London: Chatto and Windus.

Bell, D. (1973). *The Coming of Post-Industrial Society: A Venture in Social Forecasting.* New York: Basic Books.

Benevolo, L. (1971). *The Origins of Modern Town Planning.* Cambridge, MA: MIT Press.

Benjamin, W. (1969). *Illuminations: Essays and Reflections.* New York: Schocken.

Bianchi, P. (1992). "Levels of Policy and the Nature of Post-Fordist Competition", in M. Storper and A. J. Scott (eds.), *Pathways to Industrialization and Regional Development.* London: Routledge, pp. 303–315.

Bianchini, F. (1993). "Remaking European Cities: The Role of Cultural Policies", in F. Bianchini and M. Parkinson (eds.), *Remaking European Cities: The Role of Cultural Policies.* Manchester: Manchester University Press, pp. 1–20.

Binnie, J., Holloway, J., Millington, S., and Young, C. (2006). "Introduction: Grounding Cosmopolitan Urbanism: Approaches, Practices, and Policies", in J. Binnie, J. Holloway, S. Millington, and C. Young (eds.), *Cosmopolitan Urbanism.* London: Routledge, pp. 1–34.

Blackaby, F. (1978). *De-Industrialization.* London: Heinemann.

Blackley, P. R., and Greytak, D. (1986). "Comparative Advantage and Industrial Location: An Intrametropolitan Evaluation", *Urban Studies,* 23: 221–230.

Blair, H., Grey, S., and Randle, K. (2001). "Working in Film: Employment in a Project-Based Industry", *Personnel Review,* 30: 170–185.

Blau, J. R. (1989). *The Shape of Culture: A Study of Contemporary Cultural Patterns in the United States.* Cambridge: Cambridge University Press.

Bluestone, B., and Harrison, B. (1982). *The Deindustrialization of America.* New York: Basic Books.

Bobo, L. D., Oliver, M. L., Johnson, J. H., and Valenzuela, A. (2000). "Analyzing Inequality in Los Angeles", in L. D. Bobo, M. L. Oliver, J. H. Johnson, and A. Valenzuela (eds.), *Prismatic Metropolis: Inequality in Los Angeles.* New York: Russell Sage, pp. 3–50.

Boltanski, L., and Chiapello, E. (1999). *Le Nouvel esprit du capitalisme.* Paris: Gallimard.

Borjas, G. J. (2003). "The Labor Demand Curve is Downward Sloping: Reexamining the Impact of Immigration on the Labor Market", *Quarterly Journal of Economics,* 118 (4): 1335–1374.

Bourdieu, P. (1971). "Le Marché des Biens Symboliques", *L' Année sociologique,* 22: 49–126.

——(1979). *La Distinction: critique sociale du jugement.* Paris: Le Sens Commun.

Boyer, R. (1986). *La Théorie de la régulation: une analyse critique.* Paris: Algalma.

Braverman, H. (1974). *Labor and Monopoly Capitalism: The Degradation of Work in the Twentieth Century*. New York: Monthly Review Press.

Breheny, M. J., and McQuaid, R. (1987). *The Development of High Technology Industries: An International Survey*. London: Croom Helm.

Brenner, N. (2004). "Urban Governance and the Production of New State Spaces in Western Europe", *Review of International Political Economy*, 11: 447–488.

British Department of Culture Media and Sport (2001). *The Creative Industries Mapping Document*: http://www.culture.gov.uk/creative/mapping.html.

Brooks, A. C., and Kushner, R. J. (2001). "Cultural Districts and Urban Development", *International Journal of Arts Management*, 3: 4–14.

Brusco, S. (1982). "The Emilian Model: Productive Decentralization and Social Integration", *Cambridge Journal of Economics*, 6: 167–180.

Bryan, J., Hill, S., Munday, M., and Roberts, A. (2000). "Assessing the Role of the Arts and Cultural Industries in a Local Economy", *Environment and Planning A*, 32: 1391–1408.

Bunnell, T. (2002a). "Cities for Nations? Examining the City–Nation State Relation in Information Age Malaysia", *International Journal of Urban and Regional Research*, 26: 284–298.

——(2002b). "Multimedia Utopia? A Geographical Critique of High-Tech Development in Malaysia's Multimedia Supercorridor", *Antipode*, 34: 265–295.

Cairncross, F. (1997). *The Death of Distance: How the Communications Revolution Will Change Our Lives*. Boston: Harvard Business School Press.

Calenge, P. (2002). "Les Territoires de L'innovation: Les Réseaux de L'industrie de la Musique en Recomposition", *Géographie, Economie, Société*, 4: 37–56.

Carney, J., Hudson, R., and Lewis, J. (1980). *Regions in Crisis: New Perspective in European Regional Theory*. New York: St. Martin's Press.

Castells, M. (1968). "Y a-t-il une Sociologie Urbaine?", *Sociologie du travail*, 1: 72–90.

——(1972). *La Question Urbaine*. Paris: Maspéro.

——(1976). "Crise de l'état, Consommation Collective et Constradictions Urbaines", in N. Poulantzas (ed.), *La Crise de l'etat*. Paris: Presses Universitaires de France, pp. 179–208.

——(1996). *The Rise of the Network Society, Information Age; 1*. Cambridge, MA: Blackwell.

Caves, R. E. (2000). *Creative Industries: Contacts between Art and Commerce*. Cambridge, MA: Harvard University Press.

Chamberlin, E. (1933). *The Theory of Monopolistic Competition*. Cambridge, MA: Harvard University Press.

Chang, T. (2000). "Rennaissance Revisited: Singapore as a 'Global City for the Arts'", *International Journal of Urban and Regional Research*, 24: 818–831.

159

References

Cheshire, P. C. (2006). "Resurgent Cities, Urban Myths and Policy Hubris: What We Need to Know", *Urban Studies*, 43: 1231–1246.

Cochrane, A. (2007). *Understanding Urban Policy: A Critical Approach*. Oxford: Blackwell.

Coe, N. M. (2000). "On Location: American Capital and the Local Labor Market in the Vancouver Film Industry", *International Journal of Urban and Regional Research*, 24: 79–91.

Cooke, P., and Morgan, K. (1998). *The Associational Economy: Firms, Regions, and Innovation*. Oxford: Oxford University Press.

Coriat, B. (1979). *L'Atelier et le chronomètre: essai sur le Taylorisme, le Fordisme, et la production de masse*. Paris: C. Bourgois.

Cornford, J., and Robins, K. (1992). "Development Strategies in the Audio-Visual Industries: The Case of North East England", *Regional Studies*, 26: 421–436.

Courchene, T. J. (2001). "Ontario as a North American Region-State, Toronto as a Global City-Region: Responding to the NAFTA Challenge", in A. J. Scott (ed.), *Global City-Regions: Trends, Theory, Policy*. Oxford: Oxford University Press, pp. 158–190.

Currah, A. (2003). "Digital Effects in the Spatial Economy of Film", *Area*, 35: 1–10.

Currid, E. (2006). "New York as a Global Creative Hub: A Competitive Analysis of Four Theories on World Cities", *Economic Development Quarterly*, 20: 330–350.

—— (2007). *The Warhol Economy: How Fashion, Art and Music Drive New York City*. Princeton: Princeton University Press.

Daniels, P. W. (1979). *Spatial Patterns of Office Growth and Location*. New York: John Wiley.

—— (1995). "The Locational Geography of Advanced Producer Services in the United Kingdom", *Progress in Planning*, 43(Parts 2–3): 123–138.

Dear, M. J. (2000). *The Postmodern Urban Condition*. Oxford: Blackwell.

Drennan, M. P. (2002). *The Information Economy and American Cities*. Baltimore: Johns Hopkins University Press.

Duncan, O. D., Scott, W. R., Lieberson, S., Duncan, B., and Winsborough, H. H. (1960). *Metropolis and Region*. Baltimore: Johns Hopkins University Press.

Duranton, G., and Puga, D. (2004). "Micro Foundations of Urban Agglomeration Economies", in J. V. Henderson and J. F. Thisse (eds.), *Handbook of Regional and Urban Economics*, Vol. 4. Amsterdam: Elsevier, pp. 2065–2118.

Eberts, D., and Norcliffe, G. (1998). "New Forms of Artisanal Production in Toronto's Computer Animation Industry", *Geographische Zeitschrift*, 86: 120–133.

Ekinsmyth, C. (2002). "Project Organization, Embeddedness and Risk in Magazine Publishing", *Regional Studies*, 26: 229–243.

Emmanuel, A. (1969). *L'Échange inégal*. Paris: Maspéro.

References

Esser, K., Hillebrand, W., Messner, D., and Meyer-Stamer, J. (1996). *Systemic Competitiveness: New Governance Patterns for Industrial Development, GDI Book Series; No. 7.* London: Frank Cass.

Fainstein, S. S. (2001). "Inequality in Global City-Regions", in A. J. Scott (ed.), *Global City-Regions: Trends, Theory, Policy.* Oxford: Oxford University Press, pp. 285–298.

Florence, P. S. (1955). "Economic Efficiency in the Metropolis", in R. M. Fisher (ed.), *The Metropolis in Modern Life.* Garden City: Doubleday, pp. 85–124.

Florida, R. (2002). *The Rise of the Creative Class.* New York: Basic Books.

——(2004). *Cities and the Creative Class.* London: Routledge.

Frank, A. G. (1978). *Dependent Accumulation and Underdevelopment.* London: Macmillan.

Friedmann, J., and Wolff, G. (1982). "World City Formation: An Agenda for Research and Action", *International Journal of Urban and Regional Research*, 6: 309–344.

Fröbel, F., Heinrichs, J., and Kreye, O. (1980). *The New International Division of Labor.* Cambridge: Cambridge University Press.

Frost-Kumpf, H. A. (1998). *Cultural Districts: The Arts as a Strategy for Revitalizing Our Cities.* Washington, DC: Americans for the Arts.

Fuchs, G. (2002). "The Multimedia Industry: Networks and Regional Development in a Globalized Economy", *Economic and Industrial Democracy*, 23: 305–333.

García, M. I., Fernández, Y., and Zofío, J. L. (2003). "The Economic Dimension of the Culture and Leisure Industry in Spain: National, Sectoral and Regional Analysis", *Journal of Cultural Economics*, 27: 9–30.

Gereffi, G. (1995). "Global Production Systems and Third World Development", in B. Stallings (ed.), *Global Change, Regional Response: The New International Context of Development.* Cambridge: Cambridge University Press, pp. 100–142.

Gertler, M. S. (1988). "The Limits to Flexibility—Comments on the Post-Fordist Vision of Production and Its Geography", *Transactions of the Institute of British Geographers*, 13: 419–432.

Gibson, C. (2002). "Rural Transformation and Cultural Industries: Popular Music on the New South Wales Far North Coast", *Australian Geographical Studies*, 40: 337–356.

Giddens, A. (1998). *The Third Way: The Renewal of Social Democracy.* Cambridge: Polity.

Giordano, B. (2000). "Italian Regionalism or 'Padanian' Nationalism—The Political Project of the Lega Nord in Italian Politics", *Political Geography*, 19: 445–471.

Girard, M., and Stark, D. (2002). "Distributing Intelligence and Organizing Diversity in New-Media Projects", *Environment and Planning A*, 34: 1927–1944.

References

Glaeser, E., and Gottlieb, J. (2006). "Urban Resurgence and the Consumer City", *Urban Studies*, 43: 1275–1299.

Glaeser, E. L., and Maré, D. C. (2001). "Cities and Skills", *Journal of Labor Economics*, 19: 316–342.

——Kolko, J., and Saiz, A. (2001). "Consumer City", *Journal of Economic Geography*, 1: 27–50.

Glass, R. (1963). *Introduction to London: Aspects of Change*. London: Centre for Urban Studies.

Gnad, F. (2000). "Regional Promotion Strategies for the Culture Industries in the Ruhr Area", in F. Gnad and J. Siegmann (eds.), *Culture Industries in Europe: Regional Development Concepts for Private-Sector Cultural Production and Services*. Düsseldorf: Ministry for Economics and Business, Technology and Transport of the State of North Rhine-Westphalia, and the Ministry for Employment, Social Affairs and Urban Development, Culture and Sports of the State of North Rhine-Westphalia, pp. 172–177.

Goldsmith, B., and O'Regan, T. (2005). *The Film Studio*. Boulder: Rowman and Littlefield.

Gotham, K. F. (2002). "Marketing Mardi Gras: Commodification, Spectacle and the Political Economy of Tourism in New Orleans", *Urban Studies* 39: 1735–1756.

Gottdiener, M., Collins, C. C., and Dickens, D. R. (1999). *Las Vegas: The Social Production of an All-American City*. Oxford: Blackwell.

Gouldner, A. (1979). *The Future of Intellectuals and the Rise of the New Class*. New York: Seabury.

Grabher, G. (2001). "Ecologies of Creativity: The Village, the Group, and the Heterarchic Organization of the British Advertising Industry", *Environment and Planning A*, 33: 351–374.

——(2002). "Cool Projects, Boring Institutions: Temporary Collaboration in Social Context", *Regional Studies*, 36: 205–214.

——(2004). "Temporary Architectures of Learning: Knowledge Governance in Project Ecologies", *Organization Studies*, 25: 1491–1514.

Graham, B., Ashworth, G. J., and Tunbridge, J. E. (2000). *A Geography of Heritage: Power, Culture and Economy*. London: Arnold.

Graham, S., and Marvin, S. (2001). *Splintering Urbanism: Networked Infrastructures, Technological Mobilities and the Urban Condition*. London: Routledge.

Gratton, C., Dobson, N., and Shibli, S. (2001). "The Role of Major Sports Events in the Economic Regeneration of Cities", in C. Gratton and I. Henry (eds.), *Sport in the City: The Role of Sport in Economic and Social Regeneration*. London: Routledge, pp. 35–45.

Greffe, X. (2002). *Arts et artistes au miroir de l'economie*. Paris: Economica.

Hall, P. (1998). *Cities in Civilization*. New York: Pantheon.

——(2001). "Global City-Regions in the Twenty-First Century", in A. J. Scott (ed.), *Global City-Regions: Trends, Theory, Policy*. Oxford: Oxford University Press, pp. 59–77.

Hall, P. G. (1966). *The World Cities*. London: Weidenfeld and Nicolson.

Hamnett, C., and Cross, D. (1998). "Social Polarisation and Inequality in London: The Earnings Evidence", *Environment and Policy C: Government and Policy*, 16: 659–680.

Hannigan, J. (1998). *Fantasy City: Pleasure and Profit in the Postmodern Metropolis*. London: Routledge.

Hardt, M., and Negri, A. (2000). *Empire*. Cambridge, MA: Harvard University Press.

Harvey, D. (1973). *Social Justice and the City*. London: Edward Arnold.

——(1989). "From Managerialism to Entrepreneurialism—The Transformation in Urban Governance in Late Capitalism", *Geografiska Annaler, Series B—Human Geography*, 71: 3–17.

Herman, E. D., and McChesney, R. W. (1997). *The Global Media: The New Missionaries of Corporate Capitalism*. London: Cassell.

Hesmondhalgh, D. (2002). *The Cultural Industries*. London: Sage.

Heydebrand, W., and Mirón, A. (2002). "Constructing Innovativeness in New-Media Start-Up Firms", *Environment and Planning A*, 34: 1951–1984.

Hirschman, A. O. (1958). *The Strategy of Economic Development, Yale Studies in Economics 10*. New Haven: Yale University Press.

Holston, J. (2001). "Urban Citizenship and Globalization", in A. J. Scott (ed.), *Global City Regions: Trends, Theory, Policy*. Oxford: Oxford University Press, pp. 325–348.

Hong Kong Central Policy Unit. (2003). *Baseline Study on Hong Kong's Creative Industries*. Hong Kong: Centre for Cultural Policy Research, University of Hong Kong.

Hooper, B. (1998). "The Poem of Male Desires", in L. Sandercock (ed.), *Making the Invisible Visible: A Multicultural Planning History*. Berkeley: University of California Press, pp. 227–254.

Horkheimer, M., and Adorno, T. W. (1972). *Dialectic of Enlightenment*. New York: Herder and Herder.

Hoskins, C., McFadyen, S., and Finn, A. (1997). *Global Television and Film: An Introduction to the Economics of the Business*. Oxford: Clarendon Press.

Hospers, G. J. (2006). "Borders, Bridges and Branding: The Transformation of the Oresund Region into an Imagined space", *European Planning Studies*, 14: 1015–1033.

HUD. (1995). *Empowerment: A New Covenant with America's Communities. President Clinton's National Urban Policy Report*. Washington, DC: Office of Policy Development and Research, US Department of Housing and Urban Development.

References

Hudson, R. (1995). "Making Music Work? Alternative Regeneration Strategies in a Deindustrialized Locality: The Case of Derwentside", *Transactions of the Institute of British Geographers*, 20: 460–473.

Hutton, T. A. (2000). "Reconstructed Production Landscapes in the Postmodern City: Applied Design and Creative Services in the Metropolitan Core", *Urban Geography*, 21: 285–317.

——(2004). "Service Industries, Globalization, and Urban Restructuring Within the Asia-Pacific: New Development Trajectories and Planning Responses", *Progress in Planning*, 61: 1–74.

Hyman, R. (1991). "Plus ça change? The Theory of Production and the Production of Theory", in A. Pollert (ed.), *Farewell to Flexibility?* Oxford: Blackwell, pp. 259–283.

IAURIF. (2006). *Les Industries culturelles en Ile-de-France.* Paris: Institut d'Aménagement et d'Urbanisme de la Région Ile-de-France.

Indergaard, M. (2003). "The Webs They Weave: Malaysia's Multimedia Supercorridor and New York's Silicon Alley", *Urban Studies*, 40: 379–402.

Ingerson, L. (2001). "A Comparison of the Economic Contribution of Hallmark Sporting and Performing Arts Events", in C. Gratton and I. Henry (eds.), *Sport in the City: The Role of Sport in Economic and Social Regeneration.* London: Routledge, pp. 46–59.

Isin, E. F. (1999). "Citizenship, Class and the Global City", *Citizenship Studies*, 3: 267–283.

Jacobs, J. (1969). *The Economy of Cities.* New York: Random House.

Jameson, F. (1992). *Postmodernism, or, the Cultural Logic of Late Capitalism.* Durham, NC: Duke University Press.

Jayet, H. (1983). "Chômer plus souvent en région urbaine, plus souvent en région rurale", *Economie et Statistique*, 153: 47–57.

Jencks, C. (1993). *Heteropolis: Los Angeles, the Riots and the Strange Beauty of Hetero-Architecture.* London: Academy Editions: Ernst & Sohn.

Jessop, B. (2004). "Recent Societal and Urban Change: Principles of Periodization and Views on the Current Period", in T. Nielsen, N. Albertson, and P. Hemmersam (eds.), *Urban Mutations: Periodizations, Scales, and Mobilities.* Aarhus: Arkitektskolens Forlag, pp. 40–65.

Jonas, A. E. H., and Pincetl, S. (2006). "Rescaling Regions in the State: The New Regionalism in California", *Political Geography*, 25: 482–505.

Judis, J. B., and Teixeira, R. (2002). *The Emerging Democratic Majority.* New York: Scribner.

Keating, M. (2001). "Governing Cities and Regions: Territorial Restructuring in a Global Age", in A. J. Scott (ed.), *Global City-Regions: Theory, Trends, Policy.* Oxford: Oxford University Press, pp. 371–390.

Kim, S. (1995). "Expansion of Markets and the Geographic Distribution of Economic Activities: The Trends in US Manufacturing Structure, 1860–1987", *Quarterly Journal of Economics*, 110: 881–908.

References

Knox, P. L. (1995). "World Cities and the Organization of Global Space", in R. J. Johnston, P. J. Taylor, and W. J. Watts (eds.), *Geographies of Global Change*. Oxford: Blackwell, pp. 232–247.

Kong, L. (2000). "Cultural Policy in Singapore: Negotiating Economic and Socio-Cultural Agendas", *Geoforum*, 31: 409–424.

Krätke, S. (2002). "Network Analysis of Production Clusters: The Potsdam-Babelsberg Film Industry as an Example", *European Planning Studies*, 10: 27–54.

——and Taylor, P. J. (2004). "A World Geography of Global Media Cities", *European Planning Studies*, 12: 459–477.

Landry, C. (2000). *The Creative City: A Toolkit for Urban Innovators*. London: Earthscan.

Lasch, C. (1978). *The Culture of Narcissism: American Life in an Age of Diminishing Expectations*. New York: Norton.

Lash, S., and Urry, J. (1994). *Economies of Signs and Space, Theory, Culture and Society*. London; Thousand Oaks: Sage.

Lawrence, T. B., and Phillips, N. (2002). "Understanding Cultural-Products Industries", *Journal of Management Inquiry*, 11: 430–441.

Lefebvre, H. (1970). *La Révolution urbaine*. Paris: Gallimard.

——(1974). *La Production de l'espace*. Paris: Editions Anthropos.

Levy, F., and Murnane, R. J. (2004). *The New Division of Labor: How Computers are Creating the Next Job Market*. New York: Russell Sage Foundation.

Lichter, M. I., and Oliver, M. L. (2000). "Racial Differences in Labor Force Participation and Long-Term Joblessness Among Less-Educated Men", in L. D. Bobo, M. L. Oliver, J. H. Johnson, and A. Valenzuela (eds.), *Prismatic Metropolis: Inequality in Los Angeles*. New York: Russell Sage, pp. 220–248.

Lloyd, R. (2002). "Neo-Bohemia: Art and Neighborhood Development in Chicago", *Journal of Urban Affairs*, 24: 517–532.

——and Clark, T. N. (2001). "The City as an Entertainment Machine", in K. F. Gotham (ed.), *Critical Perspectives on Urban Redevelopment*. Amsterdam: JAI, Research in Urban Sociology.

Lorente, J. P. (2002). "Urban Cultural Policy and Urban Regeneration: The Special Case of Declining Port Cities in Liverpool, Marseilles, Bilbao", in D. Crane, N. Kawashima, and K. Kawasaki (eds.), *Global Culture: Media, Arts, Policy, and Globalization*. New York: Routledge, pp. 93–104.

MacLeod, G. (2001). "New Regionalism Reconsidered: Globalization and the Remaking of Political Economic Space", *International Journal of Urban and Regional Research*, 25: 804–829.

Markusen, A. (2006). "Urban Development and the Politics of a Creative Class: Evidence from a Study of Artists", *Environment and Planning A*, 10: 1921–1940.

——Hall, P., and Glasmeier, A. (1986). *High Tech America: The What, How, Where and Why of the Sunrise Industries*. Boston: Allen and Unwin.

References

Maskell, P., and Malmberg, A. (1999). "Localised Learning and Industrial Competitiveness", *Cambridge Journal of Economics*, 23: 167–185.

Massey, D. (1984). *Spatial Divisions of Labor: Social Structures and the Geography of Production*. New York: Methuen.

Massey, D. S., and Denton, N. A. (1993). *American Apartheid: Segregation and the Making of the Underclass*. Cambridge, MA: Harvard University Press.

Mattelart, A. (1976). *Multinationales et systèmes de communication: les appareils idéologiques de l'impérialisme*. Paris: Editions Anthropos.

Mayer, M. (2003). "The Onward Sweep of Social Capital: Causes and Consequences for Understanding Cities, Communities and Urban Movements", *International Journal of Urban and Regional Research*, 27: 110–132.

McDowell, L. (1999). *Gender, Identity, and Place: Understanding Feminist Geographies*. Minneapolis: University of Minneapolis Press.

——Batnitzky, A., and Dyer, S. (2007). "Division, Segmentation, and Interpellation: The Embodied Labors of Migrant Workers in a Greater London Hotel", *Economic Geography*, 83: 1–25.

McRobbie, A. (2004). "Making a Living in London's Small-Scale Creative Sector", in D. Power and A. J. Scott (eds.), *Cultural Industries and the Production of Culture*. London: Routledge, pp. 130–143.

Meade, J. E. (1952). "External Economies and Diseconomies in a Competitive Situation", *Economic Journal*, 62: 54–67.

Menger, P. M. (1993). "L'hégémonie Parisienne: économie et politique de la gravitation artistique", *Annales: Economies, Sociétés, Civilisations* No. 6: 1565–1600.

Michalet, C. A. (1987). *Le Drôle de drame du cinéma mondial*. Paris: Editions de la Découverte.

Miller, R., and Côte, M. (1987). *Growing the Next Silicon Valley: A Guide for Successful Regional Planning*. Lexington: Lexington Books.

Molotch, H. (1996). "LA as Design Product: How Art Works in a Regional Economy", in A. J. Scott and E. W. Soja (eds.), *The City: Los Angeles and Urban Theory at the End of the Twentieth Century*. Berkeley and Los Angeles: University of California Press, pp. 225–275.

——(2002). "Place in Product", *International Journal of Urban and Regional Research*, 26: 665–688.

Montgomery, J. (2007). *The New Wealth of Cities: City Dynamics and the Fifth Wave*. Aldershot: Ashgate.

Montgomery, S. S., and Robinson, M. D. (1993). "Visual Artists in New York: What's Special About Person and Place?, *Journal of Cultural Economics*, 17: 17–39.

Morris, M., and Western, B. (1999). "Inequality in Earnings at the Close of the Twentieth Century", *Annual Review of Sociology*, 25: 623–657.

Moulier Boutang, Y. (2007). *Le Capitalisme cognitif, comprendre la nouvelle grande transformation et ses enjeux*. Paris: Editions Amsterdam.

Myrdal, G. (1959). *Economic Theory and Under-Developed Regions*. London: Gerald Duckworth & Co.

Nachum, L., and Keeble, D. (2000). "Localized Clusters and the Eclectic Paradigm of FDI: Film TNC's in Central London", *Transnational Corporations*, 9: 1–37.

Neff, G., Wissinger, E., and Zukin, S. (2005). "Entrepreneurial Labor Among Cultural Producers: Cool Jobs in Hot Industries", *Social Semiotics*, 15: 307–334.

Nel, E., and Birns, T. (2002). "Place Marketing, Tourism Promotion, and Community-Based Local Economic Development in Post-Apartheid South Africa", *Urban Affairs Review*, 38: 184–208.

Norcliffe, G., and Eberts, D. (1999). "The New Artisan and Metropolitan Space: The Computer Animation Industry in Toronto", in J.-M. Fontan, J.-L. Klein, and D.-G. Tremblay (eds.), *Entre la Metropolisation et le Village Global: Les Scènes Territoriales de la Reconversion*. Québec: Presses de l'Université du Québec, pp. 215–232.

——and Rendace, O. (2003). "New Geographies of Comic Book Production in North America: The New Artisan, Distancing, and the Periodic Social Economy", *Economic Geography*, 79: 241–263.

Noyelle, T. J., and Stanback, T. M. (1984). *The Economic Transformation of American Cities*. Totowa, NJ: Rowman and Allanheld.

O'Brien, R. O. (1992). *Global Financial Integration: The End of Geography*. London: Royal Institute of International Affairs.

O'Connor, J. (1998). "Popular Culture, Cultural Intermediaries and Urban Regeneration", in T. Hall and P. Hubbard (eds.), *The Entrepreneurial City: Geographies of Politics, Regime and Representation*. Chichester: John Wiley, pp. 225–239.

OECD. (2001). *Innovative Clusters: Drivers of National Innovation Systems*. Paris: Organization for Economic Cooperation and Development.

Orfield, M. (1997). *Metropolitics: A Regional Agenda for Community and Stability*. Washington, DC and Cambridge, MA: Brookings Institution Press and the Lincoln Institute of Land Policy.

Park, R. E., Burgess, E. W., and McKenzie, R. D. (1925). *The City*. Chicago: University of Chicago Press.

Pathania-Jain, G. (2001). "Global Patents, Local Partnerships: A Value-Chain Analysis of Collaborative Strategies of Media Firms in India", *Journal of Media Economics*, 14: 169–187.

Perroux, F. (1961). *L'Économie du XXe siècle*. Paris: Presses Universitaires de France.

Philo, C., and Kearns, G. (1993). "Culture, History, Capital: A Critical Introduction to the Selling of Places", in G. Kearns and C. Philo (eds.), *Selling Places: The City as Cultural Capital, Past and Present*. Oxford: Pergamon Press, pp. 1–32.

References

Piore, M., and Sabel, C. (1984). *The Second Industrial Divide: Possibilities for Prosperity*. New York: Basic Books.

Pollard, J. (2004). "Manufacturing Culture in Birmingham's Jewellry Quarter", in D. Power and A. J. Scott (eds.), *Cultural Industries and the Production of Culture*. London: Routledge, pp. 169–187.

—— and Storper, M. (1996). "A Tale of Twelve Cities: Metropolitan Employment Change in Dynamic Industries in the 1980s", *Economic Geography*, 72: 1–22.

Pollert, A. (1991). "The Orthodoxy of Flexibility", in A. Pollert (ed.), *Farewell to Flexibility?* Oxford: Blackwell, pp. 3–31.

Power, D. (2002). "Cultural Industries in Sweden: An Assessment of Their Place in the Swedish Economy", *Economic Geography*, 78: 103–127.

—— and Scott, A. J. (2004). *Cultural Industries and the Production of Culture*. London: Routledge.

Pratt, A. C. (1997). "The Cultural Industries Production System: A Case Study of Employment Change in Britain, 1984–91", *Environment and Planning A*, 29: 1953–1974.

—— (2000). "New Media, the New Economy and New Spaces", *Geoforum*, 31: 425–436.

Putnam, R. (2000). *Bowling Alone: The Collapse and Revival of American Community*. New York: Simon and Schuster.

Rantisi, N. (2002). "The Competitive Foundations of Localized Learning and Innovation: The Case of Women's Garment Production in New York City", *Economic Geography*, 78: 441–463.

—— (2004). "The Designer in the City and the City in the Designer", in D. Power and A. J. Scott (eds.), *Cultural Industries and the Production of Culture*. London: Routledge, pp. 91–109.

Reich, R. (1992). *The Work of Nations*. New York: Vintage.

Relph, E. (1976). *Place and Placelessness*. London: Pion.

Ricardo, D. (1817). *Principles of Political Economy and Taxation*. Harmondsworth: Penguin Books (1971 edition).

Rigby, D., and Essletzbichler, J. (2005). "Technological Variety, Technological Change and a Geography of Production Techniques", *Journal of Economic Geography*, 6: 45–70.

Roost, F. (1998). "Recreating the City as Entertainment Center: The Media Industry's Role in Transforming Potsdamer Platz and Times Square", *Journal of Urban Technology*, 5: 1–21.

Roweis, S. T. (1981). "Urban Planning in Early and Late Capitalist Societies: Outline of a Theoretical Perspective", in M. Dear and A. J. Scott (eds.), *Urbanization and Urban Planning in Capitalist Society*. London: Methuen, pp. 159–177.

Royce, J. (1913). *The Problem of Christianity*, Vol. 2, *The Real World and the Christian Ideas*. New York: Macmillan.

Rullani, E. (2000). "Le Capitalisme cognitif: du déjà vu?", *Multitudes*, 2: 87–94.

Sanders, J., Nee, V., and Sernau, S. (2002). "Asian Immigrants' Reliance on Social Ties in a Multiethnic Labor Market", *Social Forces*, 81: 281–314.

Santagata, W. (2002). "Cultural Districts, Property Rights and Sustainable Economic Growth", *International Journal of Urban and Regional Research*, 26: 9–23.

Sassen, S. (1991). *The Global City: New York, London, Tokyo*. Princeton: Princeton University Press.

——(1994). *Cities in a World Economy*. Thousand Oaks: Pine Forge Press.

Sayer, A. (1989). "Postfordism in Question", *International Journal of Urban and Regional Research*, 13: 666–695.

Schmitz, H. (2007). "Regional Systems and Global Chains", in A. J. Scott and G. Garofoli (eds.), *Development on the Ground: Clusters, Networks and Regions in Emerging Economies*. London: Routledge.

Schoales, J. (2006). "Alpha Clusters: Creative Innovation in Local Economies", *Economic Development Quarterly*, 20: 162–177.

Schoenberger, E. (1989). "Thinking About Flexibility—A Response", *Transactions of the Institute of British Geographers*, 14: 98–108.

Scott, A. J. (1980). *The Urban Land Nexus and the State*. London: Pion.

——(1982). "Locational Patterns and Dynamics of Industrial Activity in the Modern Metropolis: A Review Essay", *Urban Studies*, 19: 11–142.

——(1986). "High Technology Industry and Territorial Development: The Rise of the Orange County Complex, 1955–1984", *Urban Geography*, 7: 3–45.

——(1988). *Metropolis: From the Division of Labor to Urban Form*. Berkeley: University of California Press.

——(1993). *Technopolis: High-Technology Industry and Regional Development in Southern California*. Berkeley: University of California Press.

——(1996a). "The Craft, Fashion, and Cultural Products Industries of Los Angeles: Competitive Dynamics and Policy Dilemmas in a Multi-Sectoral Image-Producing Complex", *Annals of the Association of American Geographers*, 86: 306–323.

——(1996b). "The Manufacturing Economy: Ethnic and Gender Divisions of Labor", in R. Waldinger and M. Bozorgmehr (eds.), *Ethnic Los Angeles*. New York: Russell Sage Foundation, pp. 215–244.

——(1998a). "Multimedia and Digital Visual Effects: An Emerging Local Labor Market", *Monthly Labor Review*, 121(3): 30–38.

——(1998b). *Regions and the World Economy: The Coming Shape of Global Production, Competition, and Political Order*. Oxford; New York: Oxford University Press.

——(2000a). *The Cultural Economy of Cities: Essays on the Geography of Image-Producing Industries*. London: Sage.

——(2000b). "French Cinema: Economy, Policy and Place in the Making of a Cultural Products Industry", *Theory, Culture and Society*, 17: 1–38.

References

Scott, A. J. (2005). *On Hollywood: The Place, The Industry*. Princeton: Princeton University Press.

——(2006*a*). "Entrepreneurship, Innovation and Industrial Development: Geography and the Creative Field Revisited", *Small Business Economics*, 26: 1–24.

——(2006*b*). *Geography and Economy: Three Lectures*. Oxford: Oxford University Press.

——and Pope, N. (2007). "Hollywood, Vancouver and the World: Employment Relocation and the Emergence of Satellite Production Centers in the Motion Picture Industry", *Environment and Planning A*, 39: 1364–1381.

——and Soja, E. (1996). *The City: Los Angeles and Urban Theory at the End of the Twentieth Century*. Berkeley and Los Angeles: University of California Press.

——Agnew, J., Soja, E. W., and Storper, M. (2001). "Global City-Regions", in A. J. Scott (ed.), *Global City-Regions: Trends, Theory, Policy*. Oxford: Oxford University Press, pp. 11–30.

Seaton, A. V. (1996). "Hay on Wye, the Mouse that Roared: Book Towns and Rural Tourism", *Tourism Management*, 17: 379–382.

Sennett, R. (1998). *The Corrosion of Character: The Personal Consequences of Work in the New Capitalism*. New York: W. W. Norton.

Shapiro, D., Abercrombie, N., Lash, S., and Lurry, C. (1992). "Flexible Specialisation in the Culture Industries." in H. Ernste and V. Meier (eds.), *Regional Development and Contemporary Industrial Response: Extending Flexible Specialisation*. London: Belhaven, pp. 179–194.

Simmel, G. (1903/1959). "The Metropolis and Mental Life", in K. H. Wolff (ed.), *The Sociology of Georg Simmel*. New York: Free Press, pp. 409–424.

Simon, C. J., and Nardinelli, C. (2002). "Human Capital and the Rise of American Cities, 1900–1990", *Regional Science and Urban Economics*, 32: 59–96.

Skinner, C. (2004). "The Changing Occupational Structure of Large Metropolitan Areas: Implications for the High School Educated", *Journal of Urban Affairs*, 26: 67–88.

Sklair, L. (2000). *The Transnational Capitalist Class*. Oxford: Blackwell.

Smith, N. (2002). "New globalism, New Urbanism: Gentrification as Global Urban Strategy", *Antipode*, 34: 427–450.

Soja, E. W. (2000). *Postmetropolis: Critical Studies of Cities and Regions*. Oxford: Blackwell.

STADTart. (2000). *Culture Industries in Europe. Regional Development Concepts for Private-Sector Cultural Production and Services*. Düsseldorf: Ministry of Economy and Business, Technology and Transport and Ministry for Employment, Social Affairs and Urban Development, Culture and Sport of the State of North Rhine-Westphalia.

Storper, M. (1997). *The Regional World: Territorial Development in a Global Economy, Perspectives on Economic Change*. New York: Guilford Press.

References

——and Christopherson, S. (1987). "Flexible Specialization and Regional Industrial Agglomerations: The Case of the US Motion-Picture Industry", *Annals of the Association of American Geographers*, 77: 260–282.

——and Scott, A. J. (1995). "The Wealth of Regions: Market Forces and Policy Imperatives in Local and Global Context", *Futures*, 27: 505–526.

——and Venables, A. J. (2004). "Buzz: Face-to-Face Contact and the Urban Economy", *Journal of Economic Geography*, 4: 351–370.

Swyngedouw, E. (1997). "Neither Global Nor Local: 'Glocalization' and Politics of Scale", in K. R. Cox (ed.), *Spaces of Globalization: Reasserting the Power of the Local*. New York: Guilford, pp. 137–166.

Sydow, J., and Staber, U. (2002). "The Institutional Embeddedness of Project Networks: The Case of Content Production in German Television", *Regional Studies*, (36): 215–227.

Taylor, P. J. (2005). "Leading World Cities: Empirical Evaluations of Urban Nodes in Multiple Networks", *Urban Studies*, 42: 1593–1608.

Thrift, N. (2005). *Knowing Capitalism*. London: Sage.

Throsby, D. (2001). *Economics and Culture*. Cambridge: Cambridge University Press.

Uitermark, J. (2005). "The Genesis and Evolution of Urban Policy: A Confrontation of Regulationist and Governmentality Approaches", *Political Geography*, 24: 137–163.

United Nations. (2004). *Demographic Yearbook*. New York: Department of Economic and Social Affairs, Statistical Office, United Nations.

——(2006). *Urban Agglomerations, 2005*. New York: Department of Social and Economic Affairs, Population Division, United Nations.

Ursell, G. (2000). "Television Production: Issues of Exploitation, Commodification and Subjectivity in UK Television Markets", *Media, Culture and Society*, 22: 805–825.

Van Aalst, I., and Boogaarts, I. (2002). "From Museum to Mass Entertainment: The Evolution of the Role of Museums in Cities", *European Urban and Regional Studies*, 9: 195–209.

Veltz, P. (1996). *Mondialisation, villes et territoires: l'économie d'archipel*. Paris: Presses Universitaires de France.

Vigar, G., Graham, S., and Healey, P. (2005). "In Search of the City in Spatial Strategies: Past Legacies, Future Imaginings", *Urban Studies*, 42: 1391–1410.

Waldinger, R. (2001). "The Immigrant Niche in Global City-Regions: Concept, Patterns, Controversy", in A. J. Scott (ed.), *Global City-Regions: Trends, Theory, Policy*. Oxford: Oxford University Press, pp. 299–322.

——and Bozorgmehr, M. (1996). *Ethnic Los Angeles*. New York: Russell Sage Foundation.

Wallerstein, I. (1979). *The Capitalist World Economy*. Cambridge: Cambridge University Press.

References

Ward, K. (2005). "Making Manchester Flexible: Competition and Change in the Temporary Staffing Industry", *Geoforum*, 36(2): 223–240.

Watson, S., and Gibson, K. (eds.) (1995). *Postmodern Cities and Spaces*. Oxford: Blackwell.

Whitt, J. A. (1987). "Mozart in the Metropolis: The Arts Coalition and the Urban Growth Machine", *Urban Affairs Quarterly*, 23: 15–36.

Williams, C. C. (1997). *Consumer Services and Economic Development*. London: Routledge.

Williams, R. (1976). *Keywords: A Vocabulary of Culture and Society*. London: Fontana.

Wilson, A. G. (1972). *Papers in Urban and Regional Analysis*. London: Pion.

Wilson, W. J. (1987). *The Truly Disadvantaged: The Inner City, the Underclass, and Public Policy*. Chicago: University of Chicago Press.

Wolfson, M. (1967). *Tourism Development and Economic Growth*. Paris: Organisation for Economic Development and Cooperation.

Wynne, D. (1992). "Urban Regeneration and the Arts", in D. Wynne (ed.), *The Culture Industry*. Aldershot: Avebury, pp. 94–95.

Yang, M. C., and Hsing, W. C. (2001). "Kinmen: Governing the Culture Industry City in the Changing Global Context", *Cities*, 18: 77–85.

Yau, E. C. M. (2001). "Hong Kong Cinema in a Borderless World", in E. C. M. Yau (ed.), *At Full Speed: Hong Kong Cinema in a Borderless World*. Minneapolis: University of Minnesota Press, pp. 1–28.

Young, I. M. (1999). "Residential Segregation and Differentiated Citizenship", *Citizenship Studies*, 3: 237–252.

Yun, H. A. (1999). "Multimedia and Industrial Restructuring in Singapore", in H. J. Braczyk, G. Fuchs, and H. G. Wolf (eds.), *Multimedia and Regional Economic Restructuring*. London: Routledge, pp. 376–396.

Yun, M. S. (2006). "Earnings Inequality in USA, 1969–99: Comparing Inequality Using Earnings Equations", *Review of Income and Wealth*, 1: 127–144.

Zukin, S. (1982). *Loft Living: Culture and Capital in Urban Change*. Baltimore: John Hopkins University Press.

—— (1991). *Landscapes of Power: From Detroit to Disney World*. Berkeley: University of California Press.

—— (1995). *The Cultures of Cities*. Oxford: Blackwell.

Index

Figures and notes are indexed in bold, e.g. 102**f**.

Index

Printed and bound by CPI Group (UK) Ltd, Croydon, CR0 4YY